Berichte des German Chapter of the ACM

Band 4: **Schneider, Portable Software**
Tagung I/1980 am 18. 1. 1980 in Erlangen. 176 Seiten, DM 36,–/ÖS 263,–/SFr. 32,–

Band 6: **Hauer/Seeger, Hardware für Software**
Tagung III/1980 am 10./11. 10.1980 in Konstanz. 303 Seiten, DM 54,–/ÖS 394,–/SFr. 49,–

Band 7: **Nehmer, Implementierungssprachen für nichtsequentielle Programmsysteme**
Tagung I/1981 am 20. 2. 1981 in Kaiserslautern. 208 Seiten, DM 38,–/ÖS 277,–/SFr. 34,–

Band 8: **Schlier, Personal Computing**
Tagung II/1981 am 12. 10. 1981 in Freiburg i. Br. 195 Seiten, DM 40,–/ÖS 292,–/SFr. 36,–

Band 10: **Kulisch/Ullrich, Wissenschaftliches Rechnen und Programmiersprachen**
Fachseminar am 2./3. 4. 1982 in Karlsruhe. 231 Seiten, DM 52,–/ÖS 380,–/SFr. 47,–

Band 11: **Langmaack/Schlender/Schmidt, Implementierung PASCAL-artiger
Programmiersprachen**
Tagung II/1982 am 12. 7. 1982 in Kiel. 221 Seiten, DM 46,–/ÖS 336,–/SFr. 41,–

Band 13: **Schneider, Proceedings of the International Computing Symposium 1983
on Application Systems Development**
March 22 – 24, 1983 Nürnberg. 528 Seiten, DM 90,–/ÖS 657,–/SFr. 81,–

Band 18: **Morgenbrod/Sammer, Programmierumgebungen und Compiler**
Tagung I/1984 vom 2. bis 4. 4. 1984 in München. 293 Seiten, DM 56,–/ÖS 409,–/SFr. 50,–

Band 20: **Gorny/Kilian, Computer-Software und Sachmängelhaftung**
Workshop am 29./30. 11. 1984 in Hannover. 208 Seiten. DM 48,–/ÖS 350,–/SFr. 43,–

Band 21: **Kölsch/Schmidt/Schweiggert, Wirtschaftsgut Software**
Tagung I/1985 am 26./27. 3. 1985 in Ulm. 318 Seiten, DM 58,–/ÖS 423,–/SFr. 52,–

Band 22: **Molzberger/Zemanek, Software-Entwicklung:
Kreativer Prozeß oder formales Problem?**
Seminar am 20. 3. 1985 in Neubiberg. 176 Seiten, DM 42,–/ÖS 307,–/SFr. 38,–

Band 23: **Klopcic/Marty/Rothauser, Arbeitsplatzrechner in der Unternehmung**
Tagung II/1985 am 12./13. 9. 1985 in Zürich. 355 Seiten, DM 66,–/ÖS 482,–/SFr. 59,–

Band 24: **Bullinger, Software-Ergonomie '85 Mensch-Computer-Interaktion**
Tagung III/1985 am 24./25. 9. 1985 in Stuttgart. 482 Seiten, DM 78,–/ÖS 569,– SFr. 70,–

Band 25: **Wedekind/Kratzer, Büroautomation '85**
Tagung IV/1985 vom 2. bis 4. 10. 1985 in Erlangen. 280 Seiten, DM 56,–/ÖS 409,–/SFr. 50,–

Band 27: **Remmele/Sommer, Arbeitsplätze morgen**
Tagung II/1986 vom 11. bis 14. 3. 1986 in Marburg. 431 Seiten, DM 78,–/ÖS 569,–/SFr. 70,–

Band 28: **Balzert/Heyer/Lutze, Expertensysteme '87**
Tagung I/1987 am 7./8. 4. 1987 in Nürnberg. 493 Seiten, DM 82,–/ÖS 599,–/SFr. 74,–

Band 29: **Schönpflug/Wittstock, Software-Ergonomie '87**
Tagung II/1987 vom 27. bis 29. 4. 1987 in Berlin. 512 Seiten, DM 82,–/ÖS 599,–/SFr. 74,–

Band 30: **Winkler, Proceedings of the International Workshop on Software Version
and Configuration Control**
January 27 – 29, 1988 Grassau. 478 Seiten, DM 78,–/ÖS 569,–/SFr. 70,–

Band 31: **Dillmann/Swiderski, WIMPEL '88**
Tagung I/1988 vom 28. bis 30. 6. 1988 in München. 479 Seiten, DM 78,–/ÖS 569,–/SFr. 70,–

Band 32: **Maaß/Oberquelle, Software-Ergonomie '89**
Fachtagung vom 29. bis 31. 3. 1989 in Hamburg. 509 Seiten, DM 88,–/ÖS 642,–/SFr. 79,–

Fortsetzung 3. Umschlagseite

Berichte des German Chapter
of the ACM   50

M. Sommer / W. Remmele /
K. Klöckner (Hrsg.)
Interaktion im Web –
Innovative
Kommunikationsformen

Berichte des German Chapter of the ACM

Im Auftrag des German Chapter
of the ACM herausgegeben durch den Vorstand

Chairman
Wolf-Rüdiger Gawron, BMW AG, Petuelring 130, 80788 München

Vice Chairman
Prof. Dr. Günter Riedewald, Universität Rostock, Einsteinstraße 21,
18052 Rostock

Treasurer
Eckhard Jaus, CSC Ploenzke Consulting GmbH, Zettachring 2,
70567 Stuttgart

Secretary
Roland Dürre, Interface Connection GmbH, Leipziger Straße 16,
82008 Unterhaching

Band 50

Die Reihe dient der schnellen und weiten Verbreitung neuer, für die Praxis relevanter Entwicklungen in der Informatik. Hierbei sollen alle Gebiete der Informatik sowie ihre Anwendungen angemessen berücksichtigt werden.

Bevorzugt werden in dieser Reihe die Tagungsberichte der vom German Chapter allein oder gemeinsam mit anderen Gesellschaften veranstalteten Tagungen veröffentlicht. Darüber hinaus sollen wichtige Forschungs- und Übersichtsberichte in dieser Reihe aufgenommen werden.

Aktualität und Qualität sind entscheidend für die Veröffentlichung. Die Herausgeber nehmen Manuskripte in deutscher und englischer Sprache entgegen.

Interaktion im Web – Innovative Kommunikationsformen

Herausgegeben von

Prof. Dr. rer. nat. Manfred Sommer, Universität Marburg
Werner Remmele, Siemens AG, München
Konrad Klöckner, GMD, St. Augustin

B. G. Teubner Stuttgart 1998

Die Deutsche Bibliothek – CIP-Einheitsaufnahme

Interaktion im Web – Innovative Kommunikationsformen : / hrsg. von
Manfred Sommer ... – Stuttgart : Teubner, 1998
  (Berichte des German Chapter of the ACM ; Bd. 50)
  ISBN 978-3-519-02691-4   ISBN 978-3-663-05852-6 (eBook)
  DOI 10.1007/978-3-663-05852-6

Gesamtherstellung: Präzis-Druck GmbH, Karlsruhe

Interaktion im Web

Themen der Tagung

M. Sommer
W. Remmele
K. Klöckner

Die breite Akzeptanz des *World Wide Web* zeigt, daß die ausschließlich lokale Anwendung von PC's immer mehr durch den netzorientierten Einsatz mit neuen Diensten und innovativen Kommunikationsformen ergänzt wird. Dieser Trend führt zum Ersatz einer Reihe von Techniken, wie auch zu verändertem Kommunikationsverhalten im privaten und professionellen Bereich.

Die **Fachgruppe Personal Computing der GI** hat diese Situation zum Anlaß für die Fachtagung

'Interaktion im Web - Innovative Kommunikationsformen'

genommen. Die Tagung konzentriert sich auf Themen im Bereich der Basistechnologien als auch mit deren Auswirkungen in Interaktion und Kommunikation.

Java Programming

Im Jahre 1995 wurde die Programmiersprache Java der Öffentlichkeit vorgestellt. Seitdem hat sie sich schneller verbreitet als jede andere neue Programmiersprache der letzten Jahre. Einige Ursachen für dieses Phänomen sind:

- Javaprogramme sind portabel, sie können also ohne jede Änderung auf unterschiedlichen Rechnern eingesetzt werden. Dies ist eine Voraussetzung für die Integration von Java-Anwendungen, sogenannten Applets, in Internet-Seiten.

- Javaprogramme sind für unterschiedliche Anwendungszwecke geeignet. Man kann sowohl einfache Testprogramme und Applets aber auch komplette Anwendungen erstellen, so wie sie für heutige Grafikoberflächen typisch sind.

- Trotz der interpretativen Vorgehensweise erreicht man mit Just-in-Time-Compilern effiziente Laufzeiten, die nur wenig unter denen liegen, die mit anderen Programmiersprachen erreicht werden. Moderne Prozessoren mit großen Cache-Speichern verstärken diesen Effekt.

- Java ist im Vergleich zu C++ eine wesentlich einfachere und vor allem elegantere Programmiersprache. Viele wegweisende Konzepte von Java wurden bisher noch nicht plakativ in den Vordergrund gedrängt und erschliessen sich dem Benutzer erst nach längerer eigener Anwendung.

Virtuelle Gemeinschaften

Die Web-Technologie ermöglicht den unterschiedlichsten Interessengruppen, ihre Aufgaben nicht mehr nur in lokalen Netzen, sondern auch auf dem Web zu erledigen. Diese Ausweitung der Möglichkeiten eröffnet neue Perspektiven vor allem in der Qualität der Interaktion von Workgroups. Viele Punkte sind derzeit aber noch offen und sollen im Rahmen der Tagung diskutiert werden:

- **Entwicklung gemeinsamer virtueller Arbeitsplattformen und -räume**
- **Awareness im Web**
- **Repräsentation von Personen virtuellen Arbeitsräumen**
- **Gemeinsamer Aufbau von Wissen und Information**

Real Time Kommunikation im Web

Der zunehmende Ausbau der Netze und die Möglichkeit der Reservierung von Bandbreite schaffen die Voraussetzung für Real Time Kommunikation mit Web Technologie. Einsatzszenarien wie Telefonie über das LAN sind heute nicht nur als Substitutionstechnologien zur traditionellen Telefonie realisierbar und teilweise sogar schon im Einsatz, sondern eröffnen darüber hinaus Möglichkeiten zur integrierten Kommunikation mit allen elektronischen Medien. Diese Konvergenz zwischen den traditionell getrennten Daten- und Voice-Welten führt auch zu neuen Kommunikationsformen.

- **IP als Basistechnologie für Real-Time Kommunikation.**
- **Intelligente Leistungen für die Telefonie im Web.**
- **Interaktion zwischen Kommunikation und Standard-Applikationen.**

Programmausschuß

Prof. Dr. Felix Hampe, Universität Koblenz
Prof. Dr. Wolfgang Hesse, Philipps Universität Marburg
Eckhard A. Jaus, German Chapter of the ACM
Christoph H. Hochstätter, Microsoft GmbH
Konrad Klöckner GMD, Sankt Augustin
Prof. Dr. Müller-Schloer, Universität Hannover
Werner Remmele, Siemens AG
Günter Riedewald, Universität Rostock
Gerhard Rossbach, dpunkt - Verlag für digitale Technologie GmbH
Dr. Burghardt Schallenberger, Siemens AG
Prof. Dr. Manfred Sommer - Vorsitz, Philipps Universität Marburg
Elvira Templin, Fujitsu ICL Computer GmbH

Inhaltsverzeichnis

Svend Back, Leiter Systemberatung, Sun München
„Entwicklung von Java"   11

Michael Weber, Torsten Illmann
„Using Java for the Coordination of Workflows in the
World Wide Web"   17

Uwe Egly, Gernot Koller
„JQuest: ein javabasiertes Designtool für elektronische
Fragebogen im Internet"   33

Horst Wend, Timo Salzsieder
„Vergleich JavaTM-basierter Architekturen zum Zugriff
auf relationale Datenbanken"   45

Karin Schmidt, Johannes Bumiller, Peter Manhart
„Web-Based Virtual Classrooms Supported by
Dynamic Group Building Mechanisms"   63

Marcus Ott, Carsten Huth
„Einsatz von Java Applikationen für das Organisationsdesign
virtueller Unternehmen"   73

Thomas Herrmann, Gerry Stahl
„Verschränkung von Perspektiven durch Aushandlung"   95

Thomas Koch, Wolfgang Appelt
„Gruppenwahrnehmung und Kommunikation bei Web basierten
Kooperationswerkzeugen"   113

Angi Voss, Thomas Kreifelts
„Social Construction of Knowledge"   125

Klaus H. Wolf, Konrad Froitzheim
„Benutzerraum und Dokumentenraum
                - Nachbarschaft im WWW"   137

Christoph Hochstätter, Microsoft
„Interaktion durch die Verbindung von
Telefonie-Anwendungen mit dem Web"   151

Anita Behle
„Unterstützung kooperativer Software-Wiederverwendung
durch Internet-Technologien"   153

Klement J. Fellner, Susanne Patig, Claus Rautenstrauch
„Konzeption und Entwicklung einer Internet-basierten Zeitung am Beispiel
von EIWIZ, der Elektronischen WirtschaftsInformatik-Zeitung"   165

Oliver Reiss
„Die Konvergenz von Telekooperationssystemen im Web:
Anwendungsfelder und Nutzenpotential"   181

Michael Rosemann, Boris Bachmendo
„Konzeption eines WWW-basierten Prozeßinformationssystems"   195

Dr. Wulf Dieter Bauerfeld, Deutsche Telekom Berkom GmbH
„Internet-Telefonie: Chancen und Risiken"   207

Werner Remmele, Siemens AG
„Voice over IP: Potential – Status - Trends"   215

Dieter Schinagel, Siemens AG
„IP goes MultiMedia – mediaWays, a global IP-Carrier"   221

Entwicklung von Java

Svend Back

Leiter Systemberatung, Geschäftsbereich SunSoft,
Sun Microsystems GmbH,

Zusammenfassung

Im Januar 1996 erblickte das Java Development Kit 1.0 das Licht der FTP-Server. Bereits in der Beta-Phase war erkennbar, daß sich hinter Java vielversprechende Konzepte verbergen, die einen regelrechten Run auf diese Technologie auslösten. Keine andere Programmiersprache und -plattform fand so schnell Eingang in den Anforderungskatalog für die Auswahl von IT-Personal. Wer heute Kenntnisse in Java nachweisen kann und gewillt ist, sein Wissen um Java ständig aufzufrischen, hat fast automatisch auch einen Job. Doch 900 Tage nach der Veröffentlichung ist Java noch ein gutes Stück davon entfernt, komplett standardisiert zu sein. Die Innovationsgeschwindigkeit hält alle Beteiligten auf Trab. Heute sind noch nicht alle möglichen Anwendungsbereiche ausgeleuchtet. Die Plattformunabhängigkeit und Flexibilität von Java eröffnen neue Welten. Es ist zu erwarten, daß Java neben dem Einsatz in Intranets und dem Internet besonders durch die Integration in Embedded Anwendungen und Chip-Karten eine Verbreitung finden wird, die selbst den PC-Markt als Nischenmarkt erscheinen läßt.
In dem Vortrag wird auf die Neuerungen eingegangen, die auf d e r Java Entwicklerkonferenz JavaOne in San Francisco Ende März angekündigt worden sind. Die Konferenz mit ca. 10.000 erwarteten Teilnehmern gibt wie auch im letzen Jahr die Roadmap für das nächste Jahr vor. Der Redaktionsschluß für dieses Papiers lag vor der Konferenz, so daß bei Interesse gerne auch der aktuelle Vortrag angefordert werden kann.

1 Standortbestimmung

1.1 900 Tage nach Duke

Es gibt weltweit mindestens 870 Bücher über das Thema Java[1]. Die Anzahl der Seiten, die Java referenzieren liegt bei über 2 Millionen[1]. Das Online-Verzeichnis[2] für frei zugängliche Java-Applets beinhaltet ca. 9600 registrierte WebSeiten. Laut Umfrage der Computerwoche [2] beurteilen 52% der deutschen Softwarehäuser den Einsatz von Java für 1998 als „wichtig" oder „sehr wichtig". Für den Einsatz im Jahre 1999 steigt diese Zahl auf 65 %. Es muß noch etwas dran sein an diesem Thema!
Standen 1996, in dem ersten Jahr der Existenz von Java, die Vorzüge der neuen Programmiersprache im Vordergrund und war das nächste Jahr vor allem geprägt durch eine Diskussion um Sicherheitsaspekte von Programmen im Internet. Java wurde durchweg positiv bewertet im Vergleich zum konkurrierenden Ansatz Active/X). Heute liegt der Schwerpunkt der Aktivitäten um Java auf der Vorbereitung für einen flächendeckenden Einsatz.

---

[1] nach http://www.infoseek.com und www.altavista.com, Stand Feb '98
[2] http://www.gamelan.com, Stand Feb 98

2 Programmierschnittstellen

Am Anfang stand ein minimalistischer Satz an grafischen Funktionen für die Ausstattung von grafischen Oberflächen zur Verfügung. Die Idee an sich, Applikationen über das Web zu laden, beflügelte die Programmierer. Doch die Programmierung von Applets zur Aufwertung von Internet-Seiten durch dynamische Effekte ist nicht mehr der alleinige Fokus von Java-Entwicklern.
Kommerzielle Triebfedern eröffneten neue Märkte. Es ist absehbar, daß Java durch den Einsatz z.B. in SmartCards und anderen Embedded Devices (wie z.B. SetTop-Boxen, WebPhones) schnell eine Verbreitung finden kann, die ein Vielfaches der installierten Basis an Desktop-PCs entspricht. Plattformunabhängigkeit, platzsparende Implementation und integrierte Netzwerkfähigkeit sind die Kriterien, die Java fit machen für diesen Massenmarkt. Interessanterweise war dies auch die Wiege von Java. Die Programmiersprache wurde ursprünglich entwickelt, um eine Entwicklungsumgebung für einen Personal Digital Assistant und danach für eine SetTop-Box zu realisieren.

2.1 Thin Clients und Fat Server

Doch auch im klassischen Client/Server-Umfeld und bei der wachsenden Anzahl von CORBA-basierten N-Tier Business Applikationen macht Java Fortschritte. Als Ablösung von Terminal-Lösungen bietet sich das Thin Client/Fat Server Konzept aufgrund seiner zentrale Administrationsmöglichkeit gerade zu an. Die Anforderungen an diese Applikationen sind die gleichen wie an Desktop-Programme. Mit dem Minimal-Design der frühen Java-Versionen läßt sich jedoch nicht jeder Anwender überzeugen, der den sog. „Featurismus" der Konkurrenz gewohnt ist.

2.2 Bausätze für verschiedene Einsatzgebiete

Aus den z.T. diametral unterschiedlichen Anforderungen der verschiedenen Anwendungsgebiete resultieren auch unterschiedliche Bausätze für Java-Programmierer.

- **Java Development Kit**
  Es beinhaltet den Satz an Programmierschnittstellen, die typischerweise in einem Web-Browser und auf NCs zur Verfügung stehen. Der Platzbedarf im Hauptspeicher erreicht 8MB und mehr.

- **Personal Java API**
  Es reduziert das Java Development Kit auf die Funktionen, die in Systemen (Web-Telefone, Personal Digital Assistants, Settop-Boxen) mit grafischer Oberfläche benötigt werden. Diese Geräte müssen aber nicht unbedingt Allzweckrechner darstellen. Für das Grundsystem (ohne Applikationen) sollten 2 Megabytes RAM ausreichen.

- **Embedded Java API**
  Dieser API-Satz ist für Geräte vorgesehen, die als BlackBox Funktionen übernehmen. Router, Maschinensteuerungen, Zugangskontrollsysteme wären Beispiele für den Einsatz. Für das Grundsystem (ohne Applikationen) sollten 512 Kilobytes RAM ausreichen.

- **JavaCard-API**
  Für den Einsatz in Chip-Karten wurde auch der Instruktionssatz des Java-Interpreters reduziert, um auch mit 512 Byte lokalem Speicher auszukommen.

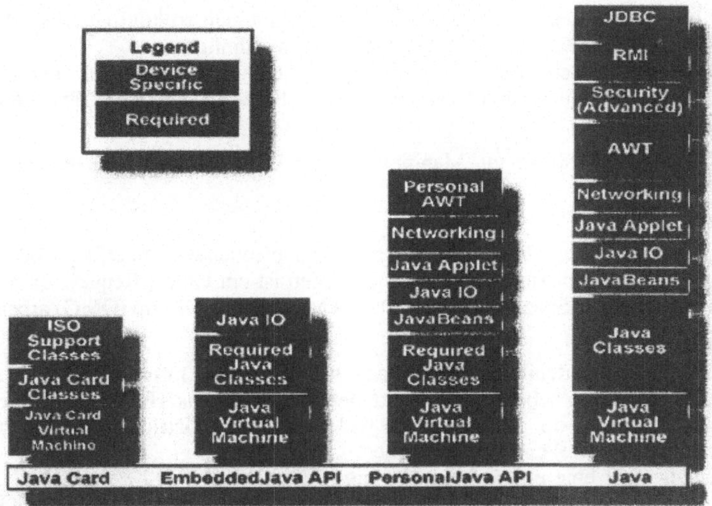

Abb. 1: Anpassung von Java-API-Sets für spezifische Anwendungsgebiete

3   Java Development Kit 1.2

Aktueller Fokus der Entwicklungsarbeit bei JavaSoft, dem Sun-Geschäftsbereich, der für Java verantwortlich zeichnet, ist eine neue Version des Java Development Kits mit neuen Funktionalitäten und/oder Verbesserungen in folgenden Bereichen:

- **JavaSecurity**
  Erweitertes Sicherheitskonzept, daß es erlaubt Java-Code mit Signaturen zu versehen und so authentifiziertem Code mehr und fein abstufbare Rechte (wie z.B. Dateizugriffe) bei der Ausführung innerhalb der Java Virtual Machine zuzugestehen. Z.B. kann es einem signierten Applet (Java-Programm in einem Web-Browser) eines Herstellers x erlaubt sein, auch direkt IP-Verbindungen zu einem System www.x.de aufzubauen, während dies allen anderen Applets verwehrt bleibt.

- **JavaSound**
  Erweiterung um Support für MIDI-, .WAV-, .AU-, und .AIFF-Dateien in CD-Qualität.

- **Remote Method Invocation (RMI)**
  RMI beschreibt die Standardkommunikation zwischen verschiedenen Java-Programmen, die sowohl beide lokal in einem Java Interpreter oder auch auf verschiedenen Systemen ablaufen können.
  RMI beherrscht ab JDK 1.2 auch Remote Object Activation.

- **Java Foundation Classes (JFC)**
  Plattformunabhängiges Fenstersystem, das neue Bedienelemente (Widgets) einführt. Gegenüber der Realisierung in der JDK-Version 1.0 und 1.1 basiert das neue Fenstersystem nicht auf den Bedienelementen des zugrundeliegenden Betriebssystems bzw. Window-Managers (z.B. MOTIF, Micosoft Foundation Classes oder QuickDraw) sondern auf 2D-Primitiven, die performant und komplett in Java realisiert sind. So ist auch auf verschiedenen Plattformen ein einheitliches GUI möglich.

Das Look & Feel der Java Foundation Classes läßt sich so konfigurieren, daß je nach Wunsch die Oberfläche Motif- oder auch Windows-ähnlich ist.
Hinzukommen noch neu Drag'n'Drop, Unterstützung für in der Bewegung eingeschränkte Benutzer und Navigation von grafischen Oberflächen über Tastatur.

- **Collections**
  Objekte und Algorithmen zur Manipulation von Listen, Stapeln und anderen abstrakten Datenmodellen.

- **JavaIDL**
  Compiler, um aus dem CORBA-IDL entsprechenden Interface-Code in Java (`idl2java`) zu generieren. Ebenfalls enthalten ist ein Object Request Broker in Java, der konform zu der Spezifikation der Object Management Group (OMG) arbeitet.

- **JavaBeans**
  Die Komponentenarchitektur JavaBeans wurde bereits mit dem JDK Version 1.1 vorgestellt. Ihr gilt neben dem Aspekt der Sicherheit das Hauptaugenmerk der SW-Entwickler, wenn sie sich für oder wider Java als Plattform (und nicht nur als Programmiersprache) entscheiden wollen.
  JavaBeans stellt eine plattformunabhängige Möglichkeit ('100% pure Java') dar, SW-Objekte standardisiert zu Applikationen zusammenzubinden. Es muss kein Quell-Code modifiziert werden, und i.d.R. überhaupt nicht zur Verfügung stehen. Grafische Programmierwerkzeuge können ohne eine Zeile Java-Code auskommen (s. Bild 2), wenn es darum geht, neue Applikationen aus vorhandenen Bausteinen zu entwerfen. Die Anzahl der verfügbaren Komponenten liegt derzeit bei 130. Sogar ein komplett in Java entwicklete Web-Browser gehört dazu [4].
  Als Basis-Technologie zum Austausch von Nachrichten zwischen den einzelnen Komponenten scheint sich die Infobus-Spezifikation von Lotus herauszukristallisieren.

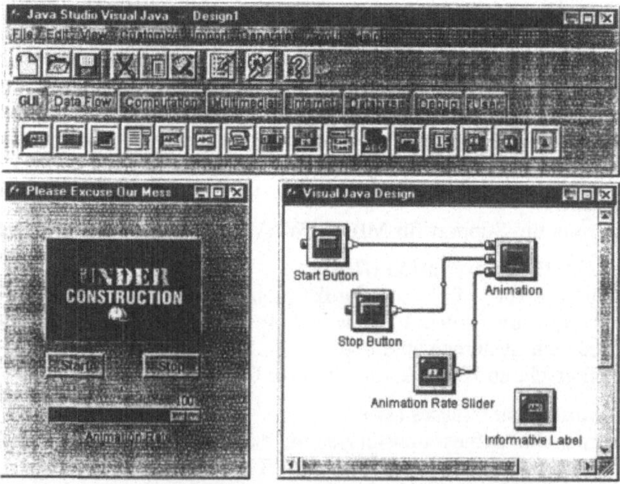

Abb. 2: Erstellung von Java-Applets (links) durch Zusammenfügen von JavaBeans (rechts)

Die Fertigstellung des Java Development Kits 1.2 [5] wird für Mitte 1998 erwartet (Stand Februar 1998: beta2).

4 Weitere Entwicklungen

Aus den Entwicklungen um Java seien hier noch ein paar Beispiele herausgegriffen, die die flächendeckenden Einsatzmöglichkeiten veranschaulichen:

4.1 Servlets

Neben der Implementation von Client-Anwendungen hat Java auch Einzug in die Programmierung von Server-Routinen gefunden. Gerade im Zusammenarbeit mit Web-Servern bieten Servlets eine Alternative zur klassischen CGI-Skript-Programmierung. Die Vorteile liegen vor allem in der besseren Skalierbarkeit, da Servlets als Erweiterung für den Web-Server zum einen nicht jedesmal neu gestartet werden müssen und zum anderen mit Multithreading-Technologie der Prozeßoverhead bei Paralleler Verarbeitung von Anfragen gering ist. Neben der Integration in bestehende Web-Server steht auch ein vollkommen in Java realisierter HTTP-Daemon (Java Web Server [6]) zur Verfügung.

4.2 Java Naming and Directory Interface (JNDI)

Je größer eine Organisation ist, um so wichtiger wird der Einsatz von zentralen Verzeichnisdiensten, um diese Informationen verwalten zu können. Ein Standard für alle Information (wie z.B. Host-Namen, Accounts, Mail-Aliases etc.) hat sich (leider) nicht durchsetzen. Mit Java geht man den Weg der Integration dieser Naming Services. Mit dem JNDI wird kein neuer Verzeichnisdienst aufgebaut, sondern bereits bestehende Dienste wie DNS, NIS oder auch LDAP zur Verfügung gestellt.

4.3 Konformität für nicht konforme Browser

Die Geschwindigkeit und die Reichhaltigkeit, mit der sich die Java-Schnittstellen entwickeln, können oder wollen nicht alle Anbieter von Web-Browsern unterstützen. Der HotJava-Browser gibt in Bezug auf Erfüllung der Java-Spezifikation die Meßlatte vor, die von den Lizenznehmer vertragsgemäß innerhalb von 6 Monaten zu implementieren ist. Diese Verzögerung stellt heute ein nicht zu unterschätzenden Bremsschuh bei der Implementation von Java-Programmen dar.

Anstelle sich jedoch auf die eingebaute Java-Funktionalität innerhalb des Netscape Navigators oder MS Internet Explorers zu verlassen, kann man durch den Einsatz des sog. Java Activators den Browser um die fehlenden Features erweitern. Als Plug-In für den jeweiligen Browser kann dessen Java-Konformität „aufgebohrt" werden [7].

4.4 Performance von Java-Anwendungen

Die Ablaufgeschwindigkeit von Java-Programmen ist durch den Einsatz von Interpretern noch nicht ganz mit der Geschwindigkeit von Programmen zu vergleichen, die in C++ oder anderen Sprachen entwickelt und dann für die jeweilige Plattform in Maschinen-Code übersetzt worden sind.

In diesem Jahr ist noch die erste Version der sog. HotSpot-Technologie zu erwarten, die adaptiv während der Ausführung von Java-Programmen, häufiger ausgeführte Programm-Sequenzen herausfiltert und zur Laufzeit in Maschinen-Code übersetzt. Bisherige Just-in-Time-Compiler übersetzten zunächst das komplette Java-Programm vor dem Start, was zu spürbaren Verzögerungen führte. Die neue Art des Just-in-Time-Compilers wird gerade diese Start-Zeiten minimieren können. Ab dem Zeitpunkt der Übersetzung ist die Ablaufgeschwindigkeit den von direkt übersetzten Programmen ebenbürtig.

4.5 ISO Standardisierung

Es ist zu erwarten, daß die sich die Innovationsgeschwindigkeit von Java zugunsten einer Standardisierung von Java verlangsamt. Die Standardisierungsgremien haben den Weg frei gemacht, daß Java ein Standard werden kann. Neu bei diesem Standardisierungsverfahren ist, daß es Sun als Erfinder von Java selbst erlaubt wird, die Vorschläge für die Spezifikation einzureichen.

Daß eine Einigung eines weltweiten Gremiums nicht über Nacht zu erwarten ist, zeigt die Standardisierungsgeschichte von C++. Für diesen Vorgänger von Java ist nach jetzt fast 10 Jahren, der „Final Draft" der ANSI-Spezifikation abgeschlossen worden. Dem erfolgreichen Einsatz von C++ hat auch der noch nicht vorhandene Standard keinen Abbruch getan. Dies gilt auch heute schon für Java.

Referenzen

[1] S. Back, S. Beier, K. Bergius, P. Majorczyk: „Professionelle Java Programmierung", ITP-Verlag, ISBN 3-8266-0249-8

[2] Computerwoche 7/98 v. 13.2.98, „Softwarehäuser setzen auf Java", Titelseite

[3] Java Studio 1.0, http://www.sun.com/studio, Download einer 30-Tage Test-Version möglich.

[4] Sun Microsystems Inc.,HotJava HTML-Komponente, http://www.javasoft.com/products/hotjava/bean/index.html, Download einer Test-Version möglich

[5] Sun Microsystems Inc., Java Development Kit, http://www.javasoft.com/products/jdk/1.2/, kostenloser Download möglich

[6] Sun Microsystems Inc.,Java Web Server, http://www.javasoft.com/products/java-server/webserver/index.html

[7] Sun Microsystems Inc.,Java Activator, http://java.sun.com/products/activator/

Adresse des Autors

Svend Back
Sun Microsystems GmbH
Bretonischer Ring 3
D-85630 Grasbrunn
Email: Svend.Back@Germany.Sun.COM

Using Java for the Coordination of Workflows in the World Wide Web

Michael Weber, Torsten Illmann
Fakultät für Informatik, Universität Ulm

Abstract

In this paper we introduce a workflow management system, called WebFlow, which is based on the world wide web and Java as its basic technologies. Java is used as the build time (modeling) language to define workflows as well as the implementation language for the run time workflow enactment. Due to the object-orientation of Java modular and extendible workflow types are possible. Modification of workflows is supported even at run time. Using WWW and Java eases the implementation effort of the workflow engine, since HTTP and the Java API already include functionality which needs not to be implemented anew. This is uploading and downloading of workflow applets and documents, authentication of clients, digital signing and especially the execution of workflows at the client site by the Java virtual machine. Thus, a very simple control server is sufficient, since the applets constituting the workflow coordinate themselves to a large extent. Webflow aims at application scenarios requiring flexible and modifiable workflows. It supports workflows which cross organizational boundaries, since it only relies on standard WWW mechanisms.

1   Introduction

Business processes which are optimized in time and flow are becoming crucial for the commercial success of companies. Such processes can be modeled as a composition of activities and subsequently mapped onto workflow management systems [1]. This phase is also called build time. During the flow of business processes the corresponding tasks constitute a so-called workflow which is controlled and coordinated by the workflow management system during run time, the second phase of workflow management. The system allocates persons to activities when the activities become active and schedules them to be performed by the individual persons. If a person receives such an activity he or she is free to choose the point in time to perform it as long as no preset time limit gets violated. If the time limit expires, the workflow management system would intervene. To perform activities a person is provided with computer-based resources, mainly documents and computer applications. These resources reside either at the person's local computer, are accessed from a server, or are sent along from person to person by the workflow management system.

A feature of most current workflow management systems is that the entire workflow is defined (modeled) before it is actually activated (instantiated). Escaping from the modeled workflow usually is impossible.

From an architectural perspective workflow management systems are mostly client/server systems. A centralized workflow engine coordinates the workflow clients being related to the workflow participants.

The growth of the internet and especially the world wide web (WWW) has caused an increasing shift from dedicated client/server software towards the exploitation of WWW technology for many kinds of applications, also for workflow management systems. Independence of operating system and hardware platform at the client side is guaranteed when using the standard WWW protocols and document formats. The basic WWW applications still follow the client/server paradigm.

However, within the WWW and also in dedicated environments the paradigm of agent-based distributed systems is evolving. Agents are active instances which are allowed to move in the network and be executed at a location different from their original host. Agents offer advantages which ideally map to the requirements of flexible, decentralized workflow management. Agents reduce network traffic, since they execute where the resources are. I.e. in a workflow scenario the necessary control takes place at the client and not through interaction with the central server. The inherent decentralization achieves higher robustness concerning network or server failures. Agent technology allows a far-reaching decoupling of workflow clients and server.

In this paper we propose a workflow management system, called WebFlow, entirely based on the WWW and Java [2] as its basic technologies. In WebFlow Java is used as the build time (modeling) language as well as the implementation language for workflow enactment. Modular and extendible workflow definitions are possible due to the object-orientation of Java. Using WWW and Java eases the implementation effort of the workflow engine, since HTTP and the Java API already include a lot of functionality, such as uploading and downloading (HTML, HTTP), authentication (HTTP), digital signatures (JDK1.1), and execution of workflows at the client site based upon the Java virtual machine [3]. Webflow aims at application scenarios requiring flexible and run time modifiable workflows which may even cross organizational boundaries.

2 Requirements and Base Technologies

Before we describe the WebFlow system in detail the core technologies being applied are sketched. Concerning workflow management we focus on the requirements such a system shall meet. Then the different design paradigms for agent systems are presented.

2.1 Workflow Management

Workflow management includes a build time and a run time dimension imposing different requirements [1] each. The build time requirements comprise all issues related to modeling of control and data flow. Most of these requirements are not specific to workflow modeling, like graphical or formal representation, ease of use, readability, abstraction and modularity, or correctness. Similar requirements are found in software engineering methodologies.

The run time requirements are related to the activation, enactment and termination of workflows. The system shall be extendible to be able to handle future unforeseen application scenarios. Workflow definitions will change over time and thus workflows should be dynamically customizable and adaptable. Continously growing application areas and increasing organizational coverage call for a scalable system. The system should be open that it can run in a distributed heterogeneous computing infrastructure.

Since in the workflow community different terminologies are used, we briefly specify the important terms being used in the paper (see also [4]).

- *Workflow:* The automation of a business process, in whole or part, during which documents, information or tasks are passed from one participant to another for action, according to a set of procedural rules.

- *Workflow definition:* A workflow definition identifies the various activities, procedural rules and associated control data used to manage the workflow during process enactment.

- *Activity:* One logical step in a workflow definition. An activity cannot be decomposed any further.

- *Composite workflow:* A workflow definition which contains activities and/or again composite workflows.

- *Workflow instance:* An instantiated workflow definition which is in some execution state (ready, executing, paused). Several instances can be derived from the same workflow definition. Workflow instances are managed by the workflow management system.

- *Subworkflow, superworkflow:* Subworkflows are used for refinement within other workflows, just as a subroutine in programming languages. Superworkflows comprise other workflows.

- *Workflow participant:* A system representation of a user which is related to the execution of an activity. This can be a representation of a human user or of a program which we call auto-user.

2.2   Agent technologies

The dominating design paradigm for distributed applications still is the client/server paradigm. Also most of the workflow management systems are implemented using this approach. With the upcoming need for network-centric computing characterized by the huge increase of internet usage and applications agent-based paradigms are evolving.

They promise to achieve much more dynamics, higher scalability, reduced network traffic, and failure tolerance through autonomy than it is possible with the client/server approach. Carzaniga et al. identify three main alternatives of agent paradigms and technologies [5].

- *Remote evaluation:* A requestor has the know-how (the code) and sends this to an executor which provides processor and I/O-resources. I.e. the client comes to the server. This paradigm is implemented in the WWW with Java servlets [6].

- *Code on demand:* A requestor has the I/O-resources and the processor. It receives the code on demand from a provider. I.e. the server comes to the client. The code is executed in the host environment until it terminates. This paradigm is implemented in the WWW using Java applets [7].

- *Mobile agent:* An agent moves to a location where the processing and I/O-resources are which it will use there on behalf of a requestor (the one sending out the agent). Mobile agents can move further on their own will without returning to the requestor. I.e. they take code and execution state with them. Odyssey [8] or aglets [9] are implementations of this paradigm.

Webflow uses the code on demand paradigm where each activity is represented by a Java applet. Opposed to most applets different Webflow applets can coordinate themselves by communicating through a shared data space.

3   Modeling of Workflows

Workflows are modeled in Java instead of using an extra definition language. This provides certain advantages:

- Workflows are defined in an object-oriented way and therefore they can be easily reused, extended and adapted to re-engineered workflows.

- There is no interpreter needed to execute workflow instances. The Java Virtual Machine (JVM) will do the interpretation.

In WebFlow, abstract workflow classes are provided which already implement the general execution procedure of a workflow. A new workflow is defined by inheriting the abstract class and implementing abstract methods defining the incoming (`initDataflow`) and outgoing data flow (`putDataflow`), the outgoing control flow (`putControlflow`) and the initialization of an application to be used for the activity (`initApplication`). The abstract methods of an activity are:

```
abstract class Workflow {
    void initApplication(Application a);
    void initDataflow();
    void putDataflow();
    void putControlflow();
    ... }
```

Jablonski and Bussler define different perspectives of a workflow [1]. The following paragraphs shortly explain how they are expressed in WebFlow.

Functional Perspective. Workflows are being developed *hierarchically*. A modeler may specify an *activity* or a *composite* workflow. While an activity is used to perform a set of simple work items for one user, a composite workflow contains several subworkflows (see figure 1).

A composite workflow contains at least two activities: a *start* and an *end* activity (see also figure 2). The start activity is called first. After its execution, subworkflows can be defined and optionally activated. When all subworkflows have been finished, the end activity is started. Its task is to evaluate the results of the subworkflows, perform hierarchical data flow and coordinate workflows which are concurrent to this composite workflow. The analogy in software development is a *divide & conquer* principle. The main routine (composite workflow) divides the problem into several subproblems (start), calls a set of subroutines to solve the subproblems (subworkflows) and finally combines the subresults to the total result (end).

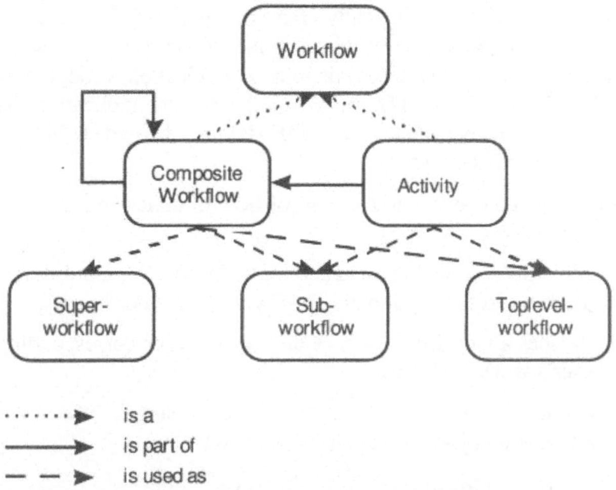

Figure 1: Relationship diagram of workflows

Defining subworkflows is done within the method `defineSubworkflows`. For instance, to specify two subworkflows „flow1" and „flow2" the following code is required:

```
void defineSubworkflows(Workflows subflows) {
    subflows.add("flow1", "MyWorkflow1.class");
    subflows.add("flow2", "MyWorkflow2.class");
}
```

„Flow1" and „flow2" are so-called workflow variables and „MyWorkflow1.class" and „MyWorkflow2.class" are the names of the real Java classes. This abstraction enables to easily change a workflow, even at run time, without modifying every subworkflow referring to it.

Operational Perspective. One main task of workflow management is the coordination of applications. In WebFlow, an activity refers to a standard application class providing a set of standard work items. These work items include reporting messages, handling dialogs, accessing databases via JDBC, uploading, downloading and modifying documents, e-mailing, etc.

To integrate external applications, the standard application class has to be inherited and the `execute` method to be overwritten. This method represents the interface from workflow to application. E.g.: the start of a local word processing program with a certain document could be implemented as follows:

```
class LocalApplication extends Application {
    Object execute (Object data) {
        String filename = (String) data;
        Runtime rt = Runtime.getRuntime();
        Process p = rt.exec("WordProcessing" + filename);
        return p;
    }
}
```

The `execute` method is called automatically after starting a workflow and enables data flow between workflow and application. Within it, one implements the access to arbitrary applications. Java easily enables the access of remote application through common protocols like RMI, RPC, CORBA and JDBC. The restriction that remote applications have to reside on the host computer is eliminated when using JDK 1.1 and trusted applets. Access to local applications also requires this feature.

Behavioral Perspective. To coordinate the control flow of concurrent workflows, three low-level primitives are provided:

`insertLink`	activates a workflow instance for a certain user and inserts a link into his work-to-do-list (optionally specifying a time constraint).
`removeLink`	finishes a workflow instance and removes the corresponding link from the user's work-to-do-list.
`updateLink`	updates a workflow instance, i.e. changes user, workflow class, application class, document or time constraint.

These primitives enable to define more complex and reusable control flow primitives like serialization, parallelism, alternative execution, and joining of parallel workflows. For example, an alternative execution may be defined as follows:

```
class MyControlFlow(Workflow w, boolean cond, Link l1, Link l2)
  {
    if (cond == true)
       w.insertLink(l1);
    else
       w.insertLink(l2);
  }
```

Since Java is an object-oriented programming language, there are no limitations to modularily designing complex high-level control flow constructs.

Informational Perspective. Since the different views of data flow and control flow often confuse, a simple method for exchanging data among workflows has been chosen in WebFlow. Figure 2 illustrates the basic idea. All concurrent workflows can read and write to a shared database. The workflow class supplies the methods `putData` and `getData` to enable this agent-based data flow. To perform hierarchical data flow, a composite workflow's start and end activity is allowed to access the database of subworkflows as well as the one of concurrent workflows. The dashed lines in figure 2 show a hierarchical data flow from subworkflows to concurrent workflows.

Organizational Perspective. The organizational perspective is designed simple and flexible. Each composite workflow has to define and assign the workflow participants being available for its subworkflows. Since definition and assignment of workflow participants is done after the execution of the start activity, the actual assignment can be done in a fixed or in a dynamic way (i.e. by querying a user's database or by prompting for them in the start activity). Subworkflows use workflow participants to coordinate workflows. This provides easy maintenance when workflow participant assignments change and offers the possibility to change workflow participants even during run time (see section 3.2).

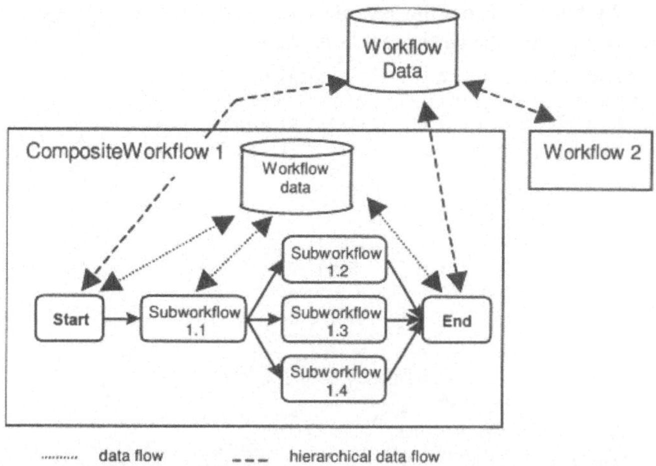

Figure 2: Data flow of composite workflows

3.1   Application Example

An example of a simple business process is explained in this section to illustrate the usability of WebFlow. Figure 3 illustrates the composite workflow „purchasing new computer equipment" which could happen in a company. It involves five workflow participants: a manager, a computer specialist, a financial specialist, a secretary and an auto-user. The scenario is as follows. The manager realizes that new computer equipment is needed. He contacts the computer specialist to get informed about possible shops and cost. He simultaneously asks the financial specialist to look for the current budget being available. If the budget is higher than the cost, the secretary will order the equipment from the chosen shop. The manager gets the message of a successfully given order. Otherwise, if the budget is too low, he gets an appropriate trouble message. Joining the parallel tasks of the specialists and the decision how to continue, is defined as an auto-user activity, i.e. it is executed automatically.

The overall composite workflow contains the start activity, the end activity and four other activities as subworkflows. Hence, the definition of four inner subworkflows reads as follows:

```
void defineSubworkflows(Workflows subflows) {
    subflows.add("budget",  "GetBudget.class");
    subflows.add("offer",   "GetOffer.class");
    subflows.add("decision","Decision.class");
    subflows.add("order",   "DoOrder.class");
}
```

Defining the workflow participants does not require the definition of the manager and the auto-user. The manager only acts in the start and end activities of the composite workflow and thus has to be defined in the superworkflow. The auto-user is a predefined workflow participant. Each workflow participant is represented by an agent which is similar to representing him by his role. From the Java perspective this means representing participants through variables which take the participant as its value.

```
void defineSubagents(Agents agents) {
    agents.add("computer-spec", "Jack");
    agents.add("financial-spec","Steve");
    agents.add("secretary",      "Susan");
}
```

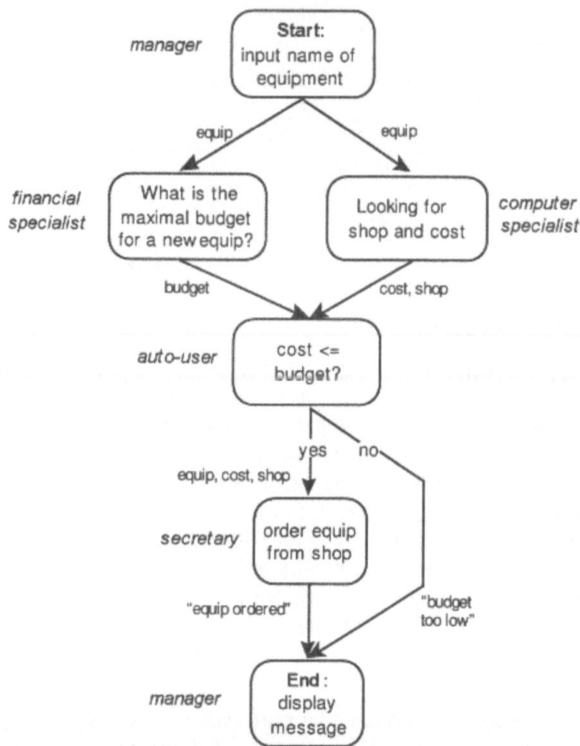

Figure 3: A sample composite workflow: "purchasing new computer equipment"

The start activity has to prompt for the required name of the equipment. The name is passed to the subworkflows within the `putDataflowStart` method. The subworkflows are activated using the `putControlflowStart` method.

```
void putDataflowStart(Dataflow subflows) {
    subflows.putData("equip", <NameOfEquipment>);
}

void putControlflowStart(Controlflow sub..) {
    subflows.insertLink("computer-spec", "offer");
```

```
        subflows.insertLink("financial-spec", "budget");
    }
```

The activity of the computer specialist has to receive as input the name of the equipment and it has to output its cost and the shop where it should be ordered. It starts the „decision" activity if necessary.

```
    class GetOffer extends Workflow {
       String equip;
       ..
       void initDataflow() {
          equip = getData("equip").toString();
       }
       void putDataflow() {
          putData("shop", <NameOfChosenShop>");
          putData("cost", <CostsOfEquipment>");
       }
       void putControlflow() {
          if (!state.isActive("Budget"))
             insertLink("auto","decision");
       }
    }
```

The „decision" activity requires the budget and cost information. Depending on whether the budget is sufficient, it activates the „order" activity or it finishes the superworkflow with outputting a failure message to the manager. Here is the code:

```
    class Decision {
       int budget, cost;
       ..
       void initDataflow() {
          budget = getData("budget").toInt();
          cost   = getData("cost").toInt();
       }

       void putControlflow() {
          if (cost < budget)
             insertLink("secretary","order");
          else                             // activate end of
superworkflow
             updateLink("super", "super");
       }

       void putDataflow() {
          if (cost > budget)
             putData("message","budget too low");
       }
    }
```

The other activities are defined in analogy to the presented ones.

3.2   Modification of Workflows

The modification of workflows is an important requirement of flexible workflow management systems, because business processes are being re-engineered permanently. In general, two different aspects may be considered. On the one hand workflow definitions can be modified, on the other hand workflow instances.

As explained previously workflow definitions are Java classes. There are two ways of modifying Java classes. Either the code is changed or it is inherited and extended. Both possibilities have advantages and disadvantages. The direct change of the source code has the advantage that all active workflow instances are dynamically being updated, too. The disadvantage is that the old workflow definition is lost and inconsistencies may be introduced easily, if either control or data flow has been changed. The more recommendable way is to inherit and extend the class. Thus, valuable parts of the old definition can be obtained and new functionality added, which allows a modular design of workflow definitions. The old definition still exists and workflow instances of both definitions can be created. The problem is that active instances are not being changed on the fly this way. The following paragraph describes a solution to this problem.

Workflow instances are instantiated workflow definitions. Related to WebFlow, instances are parameterized Java classes together with data and control data required by the instance. The instance's data consists of the names of the workflow and of the application class, the URL of the document, the user executing it, and the time constraints. The control data consists of the actual state, a unique id, the names of concurrent workflows and the concurrent workflow participants. As a special feature of WebFlow a composite workflow currently running subworkflows can control and influence the execution of these subworkflows. Workflow operations can be placed on them to change their state, i.e. (re-)activate, abort, pause, or resume them. In addition, the accompanying document, the workflow class, or the application may be exchanged on the fly. It is possible to reassign workflow participants which makes sense when users suddenly get ill or are on vacation, or when a role concept should be incorporated. These features are possible, since subworkflows are defined through workflow variables and workflow participants through agents.

4　Implementation Architecture

The key idea of the WebFlow system is the realization of workflows with Java applets. Since Java applets are locally executed programs loaded over the network together with an HTML document, they are an ideal concept for decentralized, agent-based and self-coordinating workflows. An applet is able to perform certain activities for the user (i.e. on a document), to call local and remote applications and to contact the workflow engine to invoke the subsequent control flow.

The main components of the WebFlow system are shown in figure 4. The system consists of the WebFlow engine, a database management system and a web server on the server machine. The client side comprises a web browser and local applications. Remote applications on third-party computers can be invoked as well.

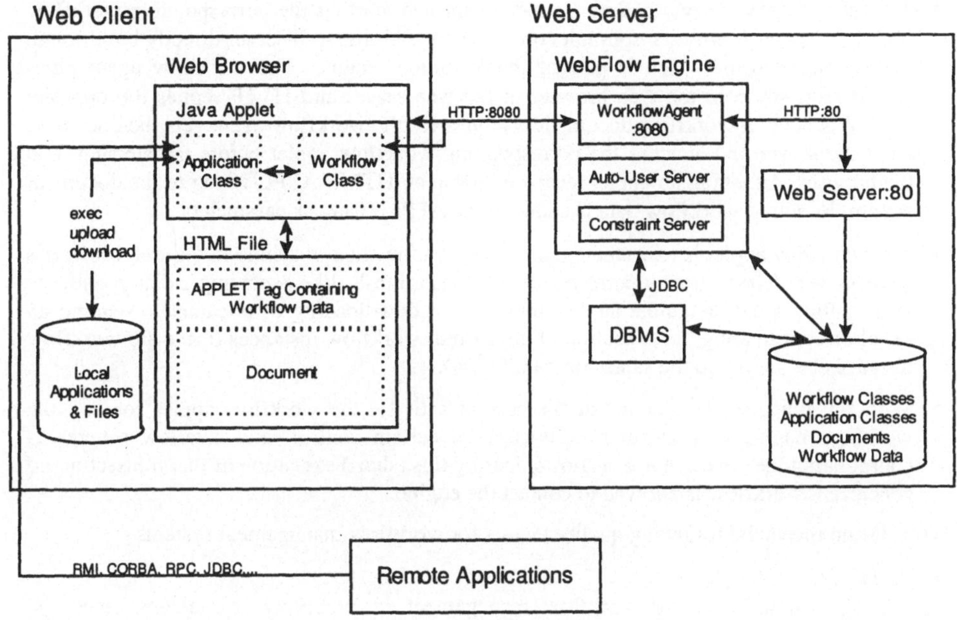

Figure 4: Architecture of the WebFlow system

To meet the quality factors and requirements of a workflow management system three major design considerations have been taken into account for the WebFlow system.

- *Decentralization*
  Each instance of a workflow activity is a self-sufficient program being executed on the client machine. The workflow engine does not act as a central control entity but executes commands of activity instances being active at client machines. The engine is only activated twice, once before starting and once after finishing an activity instance. In the first case, it supplies the activity instance with required data, in the later, it executes the control and data flow statements of the activity instance.

- *Self-Coordination*
  An activity runs independent from the engine. When finishing, it coordinates the subsequent control and data flow, i.e. it activates other activities, stops concurrent activities, or passes information to them.

The execution of a workflow instance proceeds as follows.

- The user willing to start a workflow loads his *work-to-do-list* into the web browser. The work-to-do-list is a simple HTML file containing descriptions of active workflow instances and HTML links (URLs) pointing to the instances themselves which are again HTML documents.

- Activating a link means loading the document and starting the corresponding workflow instance. The web browser does not communicate with the web server directly but through the *workflow agent* which is part of the WebFlow engine. The workflow agent pipes HTTP requests from the web browser to the web server and HTTP replies the opposite way. Whenever a workflow document is requested, the workflow agent gets the document from the server and attaches the corresponding workflow applet before transferring it to the browser. Attaching an applet means inserting an HTML APPLET tag to the document stream. Required workflow data is added to the APPLET tag as parameters.

- The workflow applet is executed automatically on the client machine in the web browser's environment. There, it may perform certain standard tasks like executing a dialog with the user, calling local or remote applications, up- or downloading of documents, sending an email or coordinating the execution of concurrent workflow instances (these are workflow instances belonging to the same superworkflow).

- With finishing the work items of the current activity, the workflow applet contacts the workflow engine to perform a transaction of certain control- and dataflow statements coordinating the concurrent workflows. During this (short) execution of that transaction no concurrent workflow is allowed to contact the engine.

This design meets the following quality factors for workflow management systems:

- *Scalability*
  The agent based style of workflow management in WebFlow is scalable, since the workflow instances are distributed on the client machines and self-coordinating. The engine simply executes commands from the workflow instances without caring about the overall process. Thus the instances on the client machines are the active part in WebFlow, the engine is passive.

- *Robustness and Correctness*
  Temporary failures of computers or the network might only cause the repetition of the lost activity in the worst case. Workflow instance and engine only connect when finishing an activity and all commands are being transmitted within a transaction. Therefore, inconsistencies do not happen due to the all-or-nothing semantics of our transactions.

- *Portability*
  Workflows and the engine may be executed on arbitrary computers and operating systems, since they are implemented entirely in Java.

- *Ease of Deployment*
  The client machines only require the installation of a web browser. Since the engine is also implemented in Java, it is portable and can be installed on the web server's machine. It communicates with every browser and server as long as they understand HTTP. Existing local and remote applications may be integrated through common protocols like RMI, RPC, CORBA, JDBC, ODBC, etc.

- *User-Friendliness*
  Dealing with workflows is simple, because they run in the user's favorite web browser and contain Java interactive components and HTML documents which have developed a standard already. The look-and-feel is similar on different machines.

* *Reusability, Extendibility and Adaptation*
  Workflow definitions are being modeled in an object-oriented and hierarchical way in Java. By inheriting workflow definitions, they can be easily extended, reused, or adapted to re-engineered business processes. Workflow instances can be dynamically adapted at run time.

4.1 Engine Architecture

One design objective of the WebFlow engine was to place as much intelligence as possible to the decentral, agent-based Java workflows and keep the intervention of the (central) engine as low as possible. Nevertheless, the engine still has to control the following tasks:

* dynamically creating work-to-do-lists when requested,

* authenticating users,

* transmitting workflow instances including actual workflow data, documents and applications to the clients.

* processing and replying to workflow requests such as `insertLink`, `removeLink`, `updateLink`, `getData`, `putData`,

* controlling time constraints of workflow instances,

* running workflows that should be executed automatically, i.e. workflows which do not need a human workflow participant.

5 WebFlow Extensions

The current WebFlow implementation is in a prototype state. The engine is fully operational as being described. Experimentations with application examples using business processes taken from the university environment have been performed as a proof of concept. These test have shown a couple of future extension possibilities.

The release of JDK1.1 [10] offers many new features most of them due to the security API and its feature of trusted applets. They are:

* digitally signing of documents (security API),

* access of local and third-party applications and databases (trusted applets),

* integrating Java Beans as workflow applications (BDK, reflection API),

* reducing transmission time of workflow applets with JAR files,

* printing of documents (improved awt API),

* uploading and downloading files without the help of the web browser (trusted applets).

The authentication scheme actually is the BASIC scheme (proposed for HTTP 1.1) which is neither a secure method of user authentication nor does it protect the transmitted entity [11]. Future versions of HTTP will provide more secure authentication schemes, e.g. digest authentication. Another possibility would be the use of *Secure HTTP, HTTP 2.0* or *SSL* which additionally guarantee a secure transmission. Such security measures are required when WebFlow shall be used in an intranet fashion within the open internet. When workflows cross organizational borders security is also a big concern.

Furthermore, our current constraint model of workflows could be improved. In the actual design, only time constraints (start date, end date, maximal duration) can be specified. A more sophisticated model requires arbitrary constraint conditions depending on time, workflow participants and workflow data.

Future work will also include the development of a Java-based, visual and object-oriented process flow editor, which allows to define workflows graphically. It should be possible to create, move and modify workflows and their the control and data flow with the mouse in a drag&drop manner. One should be able to zoom hierarchically into and out of composite workflows, e.g. by double-clicking the corresponding GUI component. In addition, every GUI component representing a workflow should provide a pop-up menu which allows to edit the properties of the workflow such as the accompanying document, application, workflow participants and constraints. The corresponding Java classes will be generated automatically.

Such an editor will push down Java from being the modeling language, as described in the paper, to an intermediate language in WebFlow. We do not see this as being contradictory to the concepts we presented. Java can still be used as a modeling language for sophisticated workflow designers. Graphical editors on top allow to have customized editors for specific application domains or for specific modeling methodologies, such as petri nets or state charts.

6   Related Work

Mobility, decentralization, agent and WWW technology have been applied to workflow management also in other systems to certain extents. We cover here a range of example systems related to the WebFlow design issues.

Mobile workflows. ProMInanD [12] follows the paradigm of migrating office folders, called electronic circulation folders (ECF). Each ECF contains in its header part among other information its migration specification. The second part holds the activity data. Since ECFs contain their routing information, they can be compared to agents migrating through a workflow management system. Local migration servers (LMS) running on each machine serve as the local run time environments for ECFs. Migration from LMS to LMS is handled by global migration servers. However, ProMInanD is based on distributed database technology and does not use any of the agent paradigms being described earlier.

WWW-based workflow clients. Many commercial and research workflow management systems use WWW browsers as a front end towards the user, e.g. [13]. As an example WWWorkflow is briefly described here. WWWorkflow [14] is designed for intranets where the users exploit WWW browsers as the only client software. WWWorkflow's architecture contains three main parts: the workfow engine, the workflow database and a CGI-gateway to the WWW. The workflow engine manages the control flow of active workflow instances in the system. It understands a „workflow language" (WFL) being a dialect of Tcl [15] augmented by commands to create, control and modify workflow instances. These workflow instances together with their current status are stored in the workflow database. The workflow engine operates on these database entries. In the database also the workflow definitions are held. Modeling constructs include steps (i.e. activities), step iterations (i.e. simple composite workflows), priorities, activation conditions and timing requirements. A common gateway interface (CGI) front end allows users to interact with the system through a triple frame to-do-list presented in their browsers. A first frame displays the actual to-do-list of a user. A middle frame presents the workspace containing a HTML document describing the activity. In the bottom frame buttons allow to submit the completion status of this activity.

Workflow agents. Dartflow [16] is a workflow management system based on mobile agents. It uses so-called process-agents which carry data and control flow of a specific workflow instance with them. Whenever an activity in the workflow instance is related to a specific machine, the process-agent migrates to this machine to perform the required actions locally, e.g. to access a local database. Process-agents are controlled by three types of agent servers: the organization server, the tracking server and the worklist server. The organization server contains generic workflow definitions and provides process-agents with the current organizational information while they are active. The organization server thus implements up-to-date role to person allocation and through this actualized routing information. Before a process-agent starts its actual workflow instance it registers at the tracking server which keeps track of all alive process-agents and their sequence of activities. The other agent servers notify the tracking server of any changes which takes appropriate action that the affected process-agents are notified before carrying out the next activity. This is possible, since all process-agents contact the worklist server after each commenced activity to add themselves to the worklist of the next user. These worklists are displayed to the users in their WWW-browser using Java applets. The work items are given as forms being handled by CGI scripts. The scripts are directly related to a process-agent responsible for the entire workflow which contains this current activity.

Decentralization. Other approaches to decentralize workflow management are also found in literature. They either build on distributed processing environments, like CodAlf [17] or BPAFrame [18], or they are based on message passing, like Exotica [19].

7 Conclusions

WebFlow is a workflow management system based on the WWW and Java as its basic technologies. Java is used as the build time (modeling) language as well as the implementation language for workflow enactment. In WebFlow modular and extendible workflow definitions are possible due to the object-orientation of Java. Inheritance of already modeled workflow defintions allows an easy construction of sophisticated workflows keeping the modeling process still manageable. Modification of workflows is possible even at run time.

Using WWW and Java eases the implementation effort of the workflow engine, since HTTP and the Java API already include functionality which has not to be implemented anew. This is uploading and downloading of workflow applets and documents, authentication of clients, digital signing and especially the execution of workflows at the client site by the Java virtual machine. I.e. a very simple control server is sufficient, since the applets constituting the workflow coordinate themselves to a large extent.

8   References

[1]   S. Jablonski, C. Bussler: Workflow Management Modeling Concepts, Árchitecture and Implementation, London, International Thomson Computer Press, 1996.

[2]   K. Arnold, J. Gosling: The Java Programming Language. ACM Press Books, Addison Wesley Longman, 1996.

[3]   T. Lindholm, F. Yellin: The Java Virtual Machine Specification. ACM Press Books, Addison Wesley Longman, 1996.

[4]   Workflow Management Coalition: Terminology & Glossary. Document No. WFMC-TC-1011, Issue 2.0, June, 1996.

[5]   A. Carzaniga, G. Picco, G. Vigna: Designing Distributed Applications using the Mobile Code Paradigms. International Conference on Software Engineering, 1997.

[6]   Sun Microsystems: The Java Servlet API White Paper. Technical report, Sun Microsystems, available from http://jserv.javasoft.com/products/java-server/webserver/beta1.0/doc/index.html, 1997.

[7]   D. Flanagan: Java in a Nutshell. O'Reilly, 1996.

[8]   J. White: Mobile Agents White Paper, available from http://www.genmagic.com/agents/ Whitepaper/whitepaper.html, 1996.

[9]   D. Lange: Programming Mobile Agents in Java - A White Paper. IBM Corp., available from http://www.ibm.co.jp/trl/aglets/whitepaper.htm, 1996.

[10]   Sun Microsystems: The Java Development Kit 1.1 Documentation. Technical report, Sun Microsystems, available from http://java.sun.com:80/products/jdk/1.1/docs/index.html, 1996.

[11]   HTTP Working Group: Hypertext Transfer Protocol - HTTP/1.1, Internet Draft, <draft-ietf-http-v11-spec-07>, 1996.

[12]   B. Karbe, N. Ramsperger, P. Weiss: Support of Cooperative Work by Electronic Circulation Folders. Conference on Office Information Systems, 1990.

[13]   C. Ellis, C. Maltzahn: The Chautauqua Workflow System. Hawaii International Conference on System Sciences, 1997.

[14]   C. Ames, S. Burleigh, S. Mitchell: WWWorkflow: World Wide Web based Workflow. Hawaii International Conference on System Sciences, 1997.

[15]   J. Ousterhout: Tcl and the Tk Toolkit. Addison-Wesley, 1994.

[16]   T. Cai, P. Gloor, S. Nog: DartFlow: A Workflow Management System on the Web using Transportable Agents. Technical Report PCS-TR96-283, Dartmouth College, 1996.

[17]   A. Schill, C. Mittasch: CodAlf: A Decentralized Workflow Management System on Top of OSF DCE and DC++. 3rd International Symposium on Autonomous Decentralized Systems, 1997.

[18]   C. Mittasch et al.: Design and Use of BPAFrame - a Decentralized CORBA-based Workflow Management System. World Computer Congress, 1996.

[19]   G. Alonso, D. Agrawal, A. Abbadi, K. Kamath, R. Günthör, C. Mohan: Exotica/FMQM: A Persistent Message-Based Architecture for Distributed Workflow Management. IFIP Working Conference on Information Systems for Decentralized Organizations, 1995.

Adresses of Authors

Prof. Dr. Michael Weber
Department for Distributed Systems
University of Ulm
D-89069 Ulm
weber@informatik.uni-ulm.de

Torsten Illmann
Department for Distributed Systems
University of Ulm
D-89069 Ulm
tillmann@hydra.informatik.uni-ulm.de

JQuest: ein javabasiertes Designtool für elektronische Fragebogen im Internet

Uwe Egly und Gernot Koller
Technische Universität Wien

Zusammenfassung

In diesem Papier stellen wir ein Designtool vor, das es erlaubt, elektronische Fragebogen einfach zu erstellen. Der Fragebogen wird dabei als Java-Applet implementiert. Statt eines Fragebogens in Papierform erhält der Befragte eine e-mail mit der Angabe einer Web-Seite, auf der der Fragebogen zu finden ist. Bei Aufruf der Seite wird der Fragebogen auf die Maschine des Befragten geladen und dort ausgeführt. Wir demonstrieren, wie das Internet neue Formen der Kommunikation ermöglicht, diskutieren wesentliche Designkriterien und beschreiben Eigenschaften der Programmiersprache Java, die für das Gelingen unseres Ansatzes notwendig sind.

1 Einleitung

Umfragen sind ein weit verbreitetes Instrument zur Ermittlung von Kundenwünschen und Kundenzufriedenheit. Beispielsweise führt die Gesellschaft für Informatik (GI) regelmässig solche Umfragen durch, um Informationen über Strukturmerkmale ihrer Mitglieder zu erhalten. Ein wesentliches Instrument solcher Umfragen stellt der *Fragebogen* dar. Die zur Zeit gängigen Umfragen nutzen zumeist lediglich den Fragebogen in Papierform. Durch die zunehmende Verbreitung von Zugangsmöglichkeiten zum Internet werden in Zukunft elektronische Umfragen möglich, wobei die technischen Möglichkeiten weit über die der traditionell organisierten "papierzentrierten" Umfragen hinausgehen werden. Beispielhaft sei hier die Möglichkeit erwähnt, Audio- oder Videosequenzen in den Fragebogen zu integrieren. Fragen wie "Wie gefällt Ihnen das folgende Videoclip" werden dann möglich. Die neuen Nutzungsmöglichkeiten werden selbst in Informatikkreisen nicht immer erkannt; die ausschließlich papierzentrierte Umfrage der GI scheint diese Ansicht zu stützen.

Wie kann man sich nun eine elektronische internetbasierte Umfrage vorstellen? Statt eines Fragebogens in Papierform erhält der potentielle Befragte eine e-mail (ähnlich dem Anschreiben zum Fragebogens in Papierform), in der er unter Angabe eines URL (*uniform resource locator*) gebeten wird, den dort abgelegten Fragebogen zu beantworten.

Der Fragebogen selbst liegt in Form eines lauffähigen Applets[1] vor, das beim Aufruf des URLs durch einen javafähigen Browser auf die Maschine des Befragten geladen und dort ausgeführt wird. Das Ausfüllen des elektronischen Fragebogens geschieht analog zum Ausfüllen des Fragebogens in Papierform, sodaß sich für den Befragten keine allzu großen Unterschiede ergeben.

In diesem Papier stellen wir ein Werkzeug vor, das es erlaubt, elektronische Fragebogen einfach zu erstellen. Diese werden dann via Internet an den Befragten versandt. Wir demonstrieren, wie das Internet neue Formen der Kommunikation ermöglicht und welche Eigenschaften der Programmiersprache Java für das Gelingen unseres Ansatzes notwendig sind. In Kapitel 2 erläutern wir Anforderungen aus der Praxis und geben eine Motivation für unseren Ansatz. Desweiteren beschreiben wir dessen Vor- und Nachteile gegenüber einer papierzentrierten Vorgehensweise. In Kapitel 3 diskutieren wir die Systemarchitektur und erläutern das Zusammenspiel der einzelnen Komponenten. Kapitel 4 gibt einen groben Überblick über die Implementierung, sowohl des Fragebogens selbst als auch des Fragebogeneditors. Mit Kapitel 5 beschließen wir den Artikel mit einer Konklusion und einem Ausblick auf zukünftige Entwicklungsmöglichkeiten.

2 Motivation und Anforderungen

In diesem Kapitel diskutieren wir praktisch orientierte Anforderungen an einen elektronischen Fragebogen sowie die sich daraus ergebenden Vor- und Nachteile. Eine Befragung gliedert sich grob in drei Schritte:

1. dem Erstellen des Fragebogens,

2. der eigentlichen Umfrage und

3. der Auswertung der Daten.

Für jeden der drei angesprochenen Schritte wird ein Werkzeug benötigt, das die Durchführung effizient unterstützt. Für den ersten Schritt ist dies der *Fragebogeneditor*, für den zweiten Schritt das Applet und das Internet und für den dritten Schritt ist dies ein Programmpaket zur statistischen Auswertung. Wir beschränken uns hier auf die Werkzeuge zur Unterstützung der Schritte eins und zwei, da sich der dritte Schritt nicht wesentlich von dem Vorgehen bei papierzentrierten Umfragen unterscheidet.

Der Fragebogen auf Papier besteht aus einer Menge von Fragen, die in beliebiger Reihenfolge beantwortet werden können. Die einzelnen Fragen besitzen naturgemäß eine fixe Reihenfolge. Zu jedem Zeitpunkt ist der komplette Fragebogen mit all seinen Fragen einsehbar. Es ist auf den ersten Blick nicht erkennbar, welche Fragen tatsächlich alle beantwortet werden müssen oder ob welche bei bestimmten Antworten entfallen. Eine wesentliche Forderung an einen elektronischen Fragebogen ist die Möglichkeit, *hierarchische*

---

[1]Dies sind (zumeist kleine) Javaprogramme, die in HTML-Seiten eingebettet sind, über das Internet geladen und auf dem Rechner des Benutzers ausgeführt werden.

Zusammenhänge zwischen Fragen zuzulassen. Desweiteren sind diese Zusammenhänge für den Befragten auch *sichtbar* zu machen.

Eng verwandt mit der erstgenannten Anforderung nach der Darstellung von Hierarchien zwischen Fragen ist die zweite Anforderung der *dynamischen Generierung eines spezialisierten Fragebogens durch Auswertung vorheriger Antworten.* Beispielsweise kann die positive Beantwortung einer Frage bewirken, daß eine Gruppe von Fragen überhaupt nicht mehr beantwortet werden muß. Der Fragebogen wird dynamisch restrukturiert und die entsprechenden Fragen werden dem Benutzer nicht zur Beantwortung präsentiert. Selbstverständlich wird die Restrukturierung zurück genommen, wenn Antworten korrigiert werden.

Eine dritte Anforderung an die dynamische Gestaltung des Fragebogens stellen *Permutationen von Fragen* bzw. *Permutationen von Antwortalternativen zu einer Frage* dar. Ein Konzept der "Umsortierung" ist nötig, um den Einfluß eines fixen Antwortverhaltens ("wähle *immer* die erste Möglichkeit") auf das Umfrageergebnis zu begrenzen.

Die vierte Anforderung ist die *dynamische Überprüfung auf zulässige Eingabewerte.* Dadurch werden unzulässige Werte von vorne herein vermieden. Die Überprüfung erstreckt sich nicht nur auf die Antworten einzelner Fragen, sondern muß auch Abhängigkeiten zwischen Fragen einbeziehen.

Eine fünfte Anforderung ist die *Unabhängigkeit des elektronischen Fragebogens von Hardware und Software* des Befragten. Da jeder Internetteilnehmer prinzipiell für eine Befragung in Frage kommt, ist diese Unabhängigkeit unbedingt erforderlich.

Werden die obigen Anforderungen an den Fragebogen erfüllt, so ergeben sich erhebliche Vorteile, sowohl unter Qualitäts- wie unter Finanzaspekten, durch den elektronischen Fragebogen. Die relevanten Vorteile werden im folgenden aufgelistet und erläutert.

++ Verbesserung der Qualität der Erhebungsdaten.

- Automatische Überwachung von Einschränkungen auf Eingabewerte.
  Als typisches Beispiel sei hier das Auswählen von mindestens n_{min} und höchstens n_{max} Antworten aus n Antwortmöglichkeiten genannt. Das Applet prüft auf eine zulässige Anzahl und läßt nur Antworten im angegebenen Bereich zu. Bei der Papierversion ist nicht zu verhindern, daß unzulässige Werte eingetragen werden.

- Permutieren von Fragen und Antworten einer Frage.
  Um den Einfluß eines starren Antwortverhaltens (z.B. "wähle *immer* die erste Möglichkeit") auf die Erhebungsdaten (statistisch) zu begrenzen, können sowohl Fragen als auch mögliche Antworten dynamisch permutiert werden. Um denselben Effekt beim Papierfragebogen zu erhalten, müssen sehr viele Permutationsvarianten erstellt werden, was u.U. die Herstellung stark verteuert.

- Automatische Generierung eines spezialisierten Fragebogens zur Laufzeit.
  Zwischen einzelnen Fragen bzw. zwischen Gruppen von Fragen bestehen oftmals starke Abhängigkeiten. Beispielsweise kann eine bestimmte Antwort auf

eine Frage die Beantwortung anderer Fragen obsolet machen ("Wenn Sie hier *ja* angegeben haben, beantworten Sie bitte Frage 5; ansonsten fahren Sie bitte mit Frage 6 fort."). Aufgrund der Antwort *ja* zur erstgenannten Frage kann nun dynamisch ein spezieller Fragebogen generiert werden, der die Frage 5 ausblendet und sofort bei Frage 6 fortfährt. Falls die Antwort zur ersten Frage verändert wird, so verändert sich selbstverständlich auch wieder der Fragebogen.

++ Kostenvorteile beim Versender.

- Es entstehen keine Druckkosten und keine Kosten für die Aussendungen (wenn man die Kosten für das Verschicken der e-mail unberücksichtigt läßt). Bei Benutzern können die Übertragungskosten für das Applet sowie die Übertragungskosten für die Antworten anfallen, sofern diese nicht durch Arbeitgeber o.ä. getragen werden.

- Die elektronische Erfassung geschieht beim Beantworter des Fragebogens. Dies verbessert nicht nur die Kostensituation durch Entfall der Datenerfassung beim Versender, sondern erhöht auch die Qualität der erfaßten Daten durch Vermeidung von Lesefehlern (bei Verwendung von Beleglesern) oder zusätzlichen Eingabefehlern bei manueller Eingabe.

Nachteilig wirken sich folgende Umstände aus.

-- Eingeschränkte Mobilität.
Der Fragebogen muß am Bildschirm beantwortet werden. Ein Mitnehmen des Fragebogens an einen anderen Ort ist zumindest erschwert.

-- Eingeschränktes Zielpublikum.
Das Zielpublikum ist z.Z. noch auf relativ wenige Haushalte mit der Möglichkeit der Nutzung des Internets eingeschränkt. Zudem sind diese Haushalte kein repräsentativer Querschnitt der Bevölkerung. Dank der rasanten Erhöhung der Zugangsberechtigungen wird die Auswirkung dieses Nachteils in Zukunft sinken.

3   Systemarchitektur

Wie oben bereits angedeutet ist das Internet (genauer das World Wide Web) das Medium, über welches der Fragebogen und die Antwortdaten versandt werden. Umgesetzt wird das System mit Hilfe der Programmiersprache Java, die sich nicht ohne Grund erstaunlich schnell als die Sprache des Internets durchgesetzt hat. Mit Java, respektive dem Java Development Kit (JDK) inklusive dem Abstract Windowing Toolkit (AWT) und verschiedenen Ergänzungspaketen steht uns eine Programmierumgebung zur Verfügung, die den unterschiedlichen Anforderungen des Systems gerecht wird. Anforderungen, die sowohl im Bereich der graphischen Benutzerschnittstelle (*user interface*), der Integration von multi-medial aufbereiteten Inhalten als auch verstärkt in Netzwerkdiensten liegen.

Es soll später noch näher darauf eingegangen werden, welche spezifischen Vorteile ein Ansatz mit Java (im Vergleich zu alternativen Ansätzen) bei der Umsetztung des elektronischen Fragebogens aufweist.

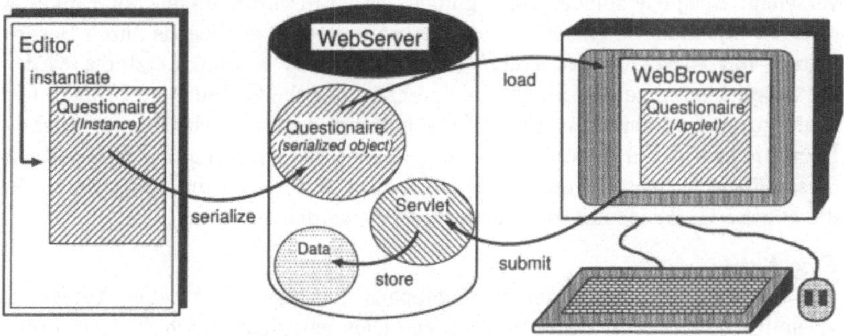

Abbildung 1: Systemarchitektur

Abbildung 1 visualisiert die Systemarchitektur einer internetbasierten Umfrage. Das dargestellte Szenario besteht im wesentlichen aus vier Komponenten: dem Fragebogeneditor, dem elektronischen Fragebogen selbst, dem Web Server und, auf der Seite des Beantworters (Client-Seite), einem javafähigen Web Browser. Bevor wir detailliert auf die wesentlichen Merkmale der unterschiedlichen Komponenten eingehen, beschreiben wir kurz deren Zusammenspiel.

Am Anfang steht die Erstellung eines Fragebogens mit dem Fragebogeneditor. Ist der Fragebogen im Dialog mit dem Benutzer erstellt, so werden die internen Datenstrukturen entsprechend den aktuellen Erfordernissen instantiiert und mittels des, von Java zur Verfügung gestellten, Serialisationsmechanismus ein lauffähiges Applet generiert. Zusätzlich wird eine Seite auf dem Web Server generiert, die den Aufruf des Applets beinhaltet.

Der zweite Schritt ist nun die eigentliche Umfrage. Nach einer freundlichen Aufforderung den unter dem URL der generierten Seite abgelegten Fragebogen zu beantworten, startet der Befragte seinen Web Browser und ruft den entsprechenden URL auf. Dadurch wird der Fragebogen auf die Maschine des Beantworters geladen und dort ausgeführt. Nach der Beantwortung des Fragebogens werden die Ergebnisse mittels eines *submit* Befehls auf den Web Server übertragen.

Wir diskutieren im folgenden die Details der einzelnen Komponenten.

- *Editor*
  Als Designtool hat der Editor die Aufgabe, das Erstellen von elektronischen Fragebogen zu vereinfachen und zu beschleunigen. Eine gute Schnittstelle zur Mensch-Maschine-Kommunikation ist hier unabdingbar. Bekannte Paradigmen wie WYSIWYG (what you see is what you get) sowie die aus gängiger Standardsoftware bekannte Copy/Paste- und Redo/Undo-Funktionaliät wird unterstützt. Verschiedene Layoutfunktionen erleichtern die Positionierung der grundsätzlich frei platzierbaren Eingabeelemente. Ebenso gute Dienste leistet ein frei skalierbares Raster,

nach dem die Elemente auf Wunsch ausgerichtet werden können. Weiters weist der Editor Eingabemöglichkeiten auf, die es dem Benutzer erlaubt, einzelne Fragen beziehungsweise einzelne Eingabeelemente mit Bedingungen (constraints) zu versehen. Beispiele solcher Bedingungen sind Einschränkungen auf Eingabewerte oder das dynamische Generieren des spezialisierten Fragebogens durch Berücksichtigung von vorher beantworteten (übergeordneter) Fragen. Letzteres wird durch die Möglichkeit erreicht, Fragen in eine hierarchische Ordnung zu bringen und mit Bedingungen zu versehen, die auf die jeweilige hierarchisch übergeordnete Frage Bezug nehmen. Zur Laufzeit des Fragebogens werden anhand der Beantwortung dieser übergeordneten Fragen, die Bedingungen ausgewertet und entschieden, ob die zugehörige Frage ein- oder ausgeblendet werden soll.

- *Fragebogen*
  Aus den im Editor entworfenen Fragebogen muß ein individuelles Applet für den aktuellen Fragebogen generiert werden. Dies geschieht durch Instantiierung der dafür vorgesehenen Klasse intern im Editor mit den Daten, die den entworfenen Fragebogen repräsentieren Das daraus resultierende Objekt wird direkt mittels Serialisation abgespeichert und auf dem Web Server abgelegt. Eingebettet in eine HTML-Seite kann dieses Applet von den Web Browsern der befragten Personen vom Web Server angefordert und ausgeführt werden. Die Aufgabe des Fragebogen-Applets ist die eigentliche Befragung durchzuführen. Dazu müssen beispielsweise die Eingaben des Benutzers unter Beachtung der definierten Bedingungen überwacht und etwaige Spezialisierungen des Fragebogens anhand der Beantwortung vorgenommen werden.

- *Web Server*
  Wie bereits erwähnt ist es die Aufgabe des Web Servers, den Web Browsern den elektronischen Fragebogen in Form eines Applets zur Verfügung zu stellen. Weiters muß er auch die nach der Beantwortung der Fragebogen anfallenden Daten entgegen nehmen und in einer Ergebnisdatei oder Datenbank ablegen. Selbstverständlich werden dabei konkurrierende Zugriffe korrekt behandelt. Als Web Server kann jedes der im Umlauf befindlichen Produkte herangezogen werden. Je nach Implementierung sollte er jedoch die Verwendung von Servlets [2, 4, 9, 10, 11, 12] gestatten.

- *Web Browser*
  Die Aufgaben des Web Browsers wurden bereits beschrieben. Er ist dafür verantwortlich, das Applet des Fragebogens vom Web Server anzufordern, es am Client-Rechner auszuführen, und die anfallenden Daten wiederum dem Web Server zukommen zu lassen.

4 Implementierung

Wie bereits erwähnt wurde der Fragebogeneditor in Java implementiert. Im folgenden wird erörtert, welche Eigenschaften der Programmiersprache von Java speziell für unser Projekt von entscheidener Bedeutung waren.

- *Netzwerkdienste*

  Für eine internetbasierte Applikation wie den elektronischen Fragebogen spielt die Unterstützung von Netzwerkdiensten natürlich eine besonders entscheidende Rolle. Java bietet hier mit dem `java.net` package eine sehr gute Unterstützung, sodaß eine Vielzahl der Netwerkoperationen nicht nur für den Endbenutzer, sondern auch schon für den Programmierer, transparent ablaufen. In diesem Bereich bringt Java möglicherweise die größte Erleichterung im Vergleich zu anderen Programiersprachen.

- *Applets*

  Wesentlich für die Implementierung des elektronischen Fragebogens in der diskutierten Form ist die Möglichkeit, Applets vom Server auf die Maschine des Befragten zu laden und dort auszuführen. Die einfachste Implemetierungsalternative wäre die Erstellung eines Fragebogens ausschließlich mittels HTML in Verbindung mit dem Common Gateway Interface (CGI). Obwohl die meisten Dialogelemente wie z.B. *checkboxes* in HTML vorhanden sind, kann eine rein HTML-basierte Implementierung, bedingt durch die "Verbindungslosigkeit" des HTTP-Protokolls, nicht überzeugen. Die sofortige Konsistenzüberprüfungen der Eingaben und vor allem die Möglichkeit, dynamisch auf Antworten des Befragten zu reagieren, wären wohl nur in sehr eingeschränkter Weise möglich. Als weitere Alternative zu in Java implementierten Applets soll der Vollständigkeit halber noch Active-X erwähnt werden. Ein Nachteil einer Lösung basierend auf Active-X ist, neben gewissen Sicherheitsproblemen, daß Active-X auf OLE und DCOM aufbaut und damit auf der 32-Bit-Windows-Plattform basiert.

  Als weitere Alternative zu in Java implementierten Applets soll der Vollständigkeit halber noch JavaScript und Active-X erwähnt werden. Bei JavaScript handelt es sich um eine in Zusammenarbeit von Netscape Communication Corporation und Sun Microsystems Inc. entstandene Scriptsprache, die direkt in den Web Browsern integriert ist. Da sich der Programmcode in die HTML-Texte einbetten läßt und die Programme auf der Clientseite direkt im Web Browser ausgeführt werden, eignet sich JavaScript sehr gut um HTML-Seiten mit zusätzlicher "Dynamik" zu versehen. Ein großes Problem stellt bei JavaScript allerdings die bestehenden Inkompatibilitäten zwischen den verschiedenen Browser-Produkten dar. Ähnlich verhält es sich mit einer Lösung basierend auf Active-X, dessen Hauptnachteil darin besteht, daß es, aufbauend auf OLE und DCOM, ausschließlich für 32-Bit-Windows-Plattformen konzipiert ist.

- *Servlets*

  Ursprünglich wurden Servlets für Suns Java Web Server entwickelt und sind nicht direkter Bestandteil des JDKs, sondern bilden ein eigenes Paket, das Java Servlet Development Kit (JSDK). Inzwischen werden Servlets von den meisten gängigen Web Servern (Apache, Netscape, Microsoft) unterstützt. Servlets bieten die Möglichkeit serverseitig Applikationen ohne graphische Oberfläche auszuführen. In unserem Fall wird dieser Mechanismus dazu benutzt, die Antwortdaten der in Umlauf gebrachten Fragebogen entgegenzunehmen und abzuspeichern. Alternativ zu Servlets könnte ein eigener Server diese Aufgabe übernehmen. Dies hätte zwar den Vorteil, völlig unabhängig vom jeweils eingesetzten Web Server zu sein, aber eben auch den Nachteil, daß zusätzlich zum Web Server ein weiterer Server aufgesetzt werden muß. Eine weitere Alternative bietet selbstverständlich das althergebrachte CGI. Dieses bietet zwar grundsätzlich die selbe Funktionalität wie Servlets, nur kann es in Verbindung mit Java zu starken Performanzeinbußen kommen, vor allem dann, wenn bei jedem CGI-Aufruf eine eigene virtuelle Maschine für Java gestartet wird.

- *Abstract Windowing Toolkit (AWT)*
  Vor allem für die Implementierung des Editors ist das Vorhandensein einer *hard- und softwareunabhängigen* Library zur Erstellung graphischer Benutzerschnittstellen (GUI-Library) von Bedeutung. Mit dem AWT ist Java auch den hohen Anforderungen des Fragebogeneditors an die graphische Benutzerschnittstelle durchaus gewachsen. Vor allem ab der Version 1.1 stellt das AWT Mechanismen bereit, die für die Implementierung des Editors in seiner jetzigen Form essentiell waren und die enorme Vereinfachungen bewirken.

  - *Serialisationsmechanismus*
    Ein ganz wichtiges Konzept ist der *Serialisationsmechanismus*, der im Editor in zweifacher Weise verwendet wird. Einerseits werden die Objekte, die die interne Reprasenatation des im Entwurf befindlichen Fragebogens darstellen, mittels dieses Mechanismus abgespeichert und wieder geladen. Andererseits wird der Mechanismus auch dazu verwendet, aus den internen Objekten den eigentlichen Fragebogen — also das Applet — zu erzeugen. Dazu wird das Applet, das den Fragebogen implementiert, intern instantiiert, mit den im Editor entworfenen Daten initialisiert und mit dem Serialisationsmechanismus als ausführbares Applet auf die Platte geschrieben. Bei der Einbettung des Applets in die HTML-Seite wird im <Applet>-Tag das `Object` Attribut anstelle des sonst gebräuchlichen `Code` Attributs benutzt.

  - *Internationalisierung*
    Die Möglichkeit, ein Programm an unterschiedliche Sprachen anzupassen, spielt für einen elektronischen Fragebogen eine bedeutende Rolle. Das AWT stellt diesbezüglich einen ganzen Satz an Mechanismen zur Verfügung. Dies beginnt bei der Bereitstellung von Fonts und Zeichensätzten für die weltweit gebräuchlichsten 25 (!) Schriften und bezieht lokale Unterschiede der alphabetischen Reihenfolge bei lexikalischen Vergleichen und der Wort- und Satzgrenzen genauso mit ein wie unterschiedliche Darstellungen von Zahlen, Dati und währungsbezogenen Beträgen. Technisch wird dies u.a. durch die Verwendung von Unicode (in der Version 2.0) [3] realisiert.

Allerdings ist auch zu bemerken, daß das AWT im direkten Vergleich mit ähnlichen Produkten, wie sie beispielsweise für C++ existieren, (noch) deutliche Schwächen zeigt. Ein Grund dafür ist, daß das AWT noch mitten in der Entwicklung steckt. Kaum ein anderer Bereich der gesamten Programmierumgebung für Java hat beim

Versionssprung von 1.0 auf die Version 1.1 so viele, teilweise grundlegende, Änderungen erfahren. Eine ähnlich große Umstellung ist beim Übergang von der Version 1.1 auf die Version 1.2 zu erwarten. Vor allem die Integration des Swing-Paketes wird hoffentlich eine Vielzahl an Problemen beseitigen, die in Zusammenhang mit den sogenannten *native peers* der Komponenten des AWT stehen.

- *Multi-Media*
  Das Einbeziehen von multimedialen Inhalten in Fragebogen stellt eine entscheidende Erweiterung der diesbezüglich beschränkten Möglichkeiten einer "papierzentrierten" Umfrage dar. In der derzeitigen Version des AWT gibt es allerdings nur spärliche Unterstützung, solche multimedialen Inhalte in eine Applikation zu integrieren. Doch liegt bereits seit einiger Zeit das Java Media Framework API (JMF) vor, das diesem Manko Abhilfe bereiten soll. Zwar befindet sich das JMF noch im Beta-Stadium, doch werden bereits die meisten Multimediastandards wie QuickTime, MPEG-1, MPEG-2, AVI, MIDI, AU, WAV und AIFF unterstützt.

5 Konklusion und Ausblick

Wir haben demonstriert, wie das Internet und insbesondere Eigenschaften der Programmiersprache Java einem eleganten Systemansatz für ein Werkzeug zur Erstellung elektronischer Fragebogen ermöglicht. Gegenüber herkömmlichen Umfragen, die zumeist auf Fragebogen in Papierform basieren, bietet die elektronische Variante erhebliche Vorteile. Neben den geringeren Kosten für die Erstellung ist insbesondere die verbesserte Qualität der Umfragedaten hervorzuheben. Voraussetzung für die Qualitätssteigerung ist die Möglichkeit, dynamisch Überprüfungen bzw. Änderungen des Fragebogens vorzunehmen.

Durch die Verwendung von Internettechnologien werden auch erweiterte Formen der Kommunikation möglich. Die Fähigkeit des Abspielens von Sound- und Videodateien werden in nächster Zukunft zu einer Verbreiterung des Einsatzgebietes solcher elektronischen Fragebogen führen. Es soll an dieser Stelle aber nicht verschwiegen werden, daß es bei der Nutzung dieser neuen Möglichkeiten, insbesondere bei Verwendung schmalbandiger Übertragungsmedien wie Telefonleitungen und Modems, zu erheblichen Geschwindigkeitseinbußen kommen kann, da sowohl Sound- als auch Videoaufzeichnungen erhebliche Speichergrößen beanspruchen. Wenn ausgiebig von Sound- und Videoaufzeichnungen Gebrauch gemacht wird ist ein Versand dieser Dateien auf einer CD vorstellbar. Nur der Fragebogen *ohne* die entsprechenden Dateien sowie die Antworten des Befragten werden dann noch via Internet versandt.

Literatur

[1] Mary Campione and Katy Walrath. *The Java Tutorial: Object-Oriented Programming for the Internet.* Addison-Wesley, 1996.

[2] Phil Inje Chang. Inside the Java Web Serer: An Overview of Java Web Server 1.0, Java Servlets, and the JavaServer Architecture, August 1997. Elektronisches Manuskript unter http://java.sun.com/features/1997/aug/jws1.html.

[3] The Unicode Consortium. *The Unicode Standard, Version 2.0*. Addison-Wesley, 1996.

[4] Dale Dougherty. Understanding Java Servlets, Oktober 1997. Elektronisches Manuskript unter http://webreview.com/97/10/10/feature/colton.html.

[5] David Flanagan. *Java in a Nutshell*. O'Reilly, second edition, May 1997.

[6] James Gosling and Henry McGilton. The Java Language Environment: A White Paper, Mai 1996. Elektronisches Manuskript unter http://java.sun.com/docs/white/langenv/.

[7] Alexander Grosse. Tor zum Virtuellen: Javas grafische Fähigkeiten: 2D-/3D-API. *iX - Magazin für professionelle Informationstechnik*, 10:136–141, Oktober 1997.

[8] Elliotte Rusty Harold. *Java Network Programming*. O'Reilly, 1997.

[9] Jörn Heid. Kettenreaktion: Servlets als CGI: serverseitige Java-Anwendung. *iX - Magazin für professionelle Informationstechnik*, 11:166–171, November 1997.

[10] JavaSoft. Introduction to Servlets, 1997. Elektronisches Manuskript unter http://jserv.javasoft.com/products/java-server/documentation/toolkit1.1beta/servlets/intro.html.

[11] Rich Kadel. Java Servlets in Netscape Enterprise Server, Dezember 1997. Elektronisches Manuskript unter http://developer.netscape.com/news/viewsource/kadel_servlet.html.

[12] Rich Kadel. Server-Side Java: Extending Web Servers through Java, Juli 1997. Elektronisches Manuskript unter http://developer.netscape.com/news/viewsource/kadel_ssjava.html.

[13] Helmut F. Reibold, Ulrich Schmitz, and Jürgen Seeger. Die Rivalen: Was ActiveX von Java, JavaScript und Plugins unterscheidet. *iX - Multiuser Multitasking Magazin*, 2:108–111, Februar 1997.

[14] Lisa Stapleton. How to Create a Multimedia Player: Using Java Media, November 1997. Elektronisches Manuskript unter http://developer.javasoft.com/developer/javaInDepth/mediaplayer.html.

[15] Uwe Steinmüller. Let it swing. *iX - Magazin für professionelle Informationstechnik*, 10:132–135, Oktober 1997.

[16] Sun Microsystems, Inc. JAVA 2D API - Enhanced Graphics and Imaging for Java - White Paper, 1997. Elektronisches Manuskript unter http://java.sun.com/marketing/collateral/2d.html.

[17] Sun Microsystems, Inc. Java Media Player - White Paper, 1997. Elektronisches Manuskript unter http://java.sun.com/marketing/collateral/jmf.html.

[18] Greg Voss. JavaServer Technologies, part I, November 1997. Elektronisches Manuskript unter http://developer.javasoft.com/developer/javaInDepth/java-server/java-server.html.

[19] Greg Voss. JavaServer Technologies, Part II, November 1997. Elektronisches Manuskript unter http://developer.javasoft.com/developer/javaInDepth/java-server/javaserver2.html.

Adressen der Autoren

Dr. Uwe Egly
Abt. Wissensbasierte Systeme 184/3
Technische Universität Wien
Treitlstraße 3
A–1040 Wien
e-mail: uwe@kr.tuwien.ac.at

Gernot Koller

Schubertgasse 24
A-2340 Mödling
e-mail: gernot@kr.tuwien.ac.at

Aktuelle Informationen zu JQuest finden Sie unter http://www.kr.tuwien.ac.at/~JQuest

Vergleich Java™-basierter Architekturen zum Zugriff auf relationale Datenbanken

Horst Wend und Timo Salzsieder[1]

Oracle Deutschland GmbH, Consulting Services

Zusammenfassung

Die Programmiersprache Java hat sich im Laufe der Zeit zur Schlüsseltechnologie bei der Entwicklung Internet/Intranet-basierter Applikationen entwickelt. Java verspricht portable Anwendungen, die in low-cost, netzwerk-orientierten Umgebungen ablaufen können. In diesem Zusammenhang spielen effiziente Datenbankzugriffe im professionellen Umfeld eine wichtige Rolle. Das vorliegende Paper untersucht unterschiedliche Lösungsansätze in Bezug auf Performanz und Implementierungsaufwand. Hierzu beleuchten wir Java Database Connectivity (JDBC™), in Java eingebundenes SQL (embedded SQL, SQLJ) und einen auf CORBA aufbauenden Ansatz. Neben diesen rein technischen Betrachtungen werden auch zukünftig zu erwartende Tendenzen in diesem Bereich behandelt.

1. Einführung

"Das Netzwerk ist der Computer" lautet ein vielzitierter Ausspruch von Scott McNealy, dem Chef des kalifornischen Unternehmens Sun Microsystems. Dieser Ausspruch hat durch Java, CORBA und Netzwerk Computing eine ganz neue Bedeutung erlangt.

Konzentrierte sich in den letzten Jahren ein Großteil der Entwicklung von Client/Server-Anwendungen auf die Trennung von Daten und denjenigen Programmteilen, die auf diese Daten zugreifen, verfolgt man heute häufig einen neues Architekturmodell: das Drei-Schichten-Modell (3-Tier Modell). Diese drei Schichten setzten sich aus den eigentlichen Benutzeranwendungen , dem Datenbanksystem und der Mittelschicht - dem Programm zur Umsetzung der eigentlichen Geschäftslogik - zusammen.

Im vorliegende Paper werden verschiedene Lösungsansätze basierend auf unterschiedli-chen Architekturmodellen untersucht, insbesondere in Bezug auf Performanz und Implementierungsaufwand.

Kapitel 2 geht zunächst kurz auf die verschiedenen JDBC-Treiber, auf einen auf JDBC aufbauenden Java Precompiler (SQLJ) und auf CORBA näher ein. Kapitel 3 stellt die verschiedenen Architekturmodelle vor und in Kapitel 4 werden die Performanz Gesichtspunkte untersucht. Mit einem Fazit schließt dieses Paper ab.

---

1   Anmerkung: Die vorliegende Darstellung repräsentiert nicht unmittelbar die der Oracle Deutschland GmbH oder der Oracle Corporation.

2. JDBC, SQLJ und CORBA

2.1 Java Database Connectivity (JDBC)

Eine der wesentlichen Begleiterscheinungen von Java ist die Bereitschaft aller Beteiligten, von vornherein Standards für Klassenbibliotheken und Spracherweiterungen festzulegen. Zur Speicherung von Daten sowie für die Datenbankanbindung stehen im Java Development Kit JDK 1.1 zwei standardisierte APIs (Application Programming Interfaces) zur Verfügung:

- JDBC (Java Database Connectivity) und
- Object Serialization.

Das Object Serialization API ist zwar sehr eng in Java integriert, bietet aber keine echte Datenbank-Funktionalität wie z.B. Transaktionssicherheit oder Mehrbenutzerzugriff. Für diese Anforderungen wurde JDBC entwickelt, das den Zugriff auf relationale Datenbanken ermöglicht. Java-Entwicklern steht damit eine Schnittstelle zur Verfügung, mit der sich Anwendungen und Werkzeuge entwickeln lassen, die auf unterschiedliche Datenbanken auf einheitliche Weise zugreifen können. JDBC selbst basiert auf dem X/Open-SQL-Call-Level-Interface.

Wie Abbildung 1 verdeutlicht besteht JDBC selbst aus zwei Schichten, dem JDBC-API und dem JDBC-Treiber API. Letztere Schicht stellt die Kommunikation zwischen dem JDBC-Manager und den verschiedenen JDBC-Treibern sicher. Die JDBC-API definiert Java-Klassen und Schnittstellen (Interfaces), die Datenbankverbindungen, SQL-Befehle, Ergebnismengen, Datenbank-Metadaten etc. repräsentieren. Damit wird Java-Entwicklern die Möglichkeit geboten aus Java heraus SQL-Befehle abzusetzen und die daraus resultierenden Ergebnisse weiter zu verarbeiten.

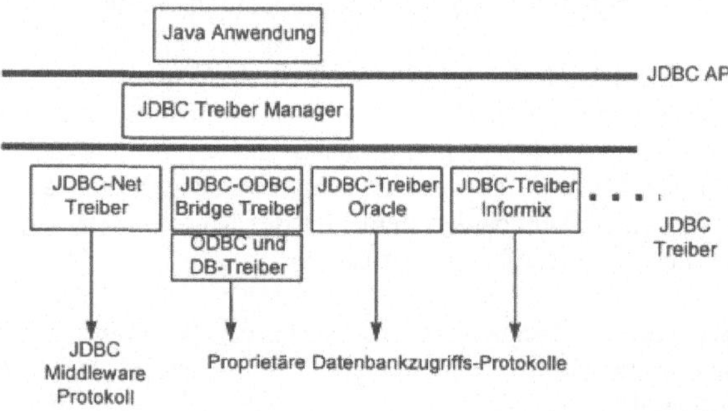

Abbildung 1: Aufbau von JDBC

Der JDBC Treiber Manager selbst enthält nur wenige Methoden. Die meisten Datenbankaufrufe gehen direkt von den Java Programmen an die entsprechenden Treiber. JDBC-Treiber können entweder vollständig in Java selbst geschrieben sein, so daß sie z.B. als Teil eines Applets geladen werden können, oder sie können unter Verwendung herstellereigener Datenbank-APIs entwickelt werden. Die daraus resultierenden verschiedenen JDBC-Treiber werden in vier verschiedene Typen eingeteilt [23].

2.1.1 JDBC-ODBC bridge driver (Type 1)

Die von Sun und Intersolv entwickelte JDBC-ODBC Brücke setzt auf bestehende ODBC-Treiber auf. Sie ermöglicht damit JDBC-Datenbankzugriffe via ODBC-Treiber. Bei ODBC handelt es sich um eine herstellerspezifische C-Schnittstelle, die wiederum auf dem herstellerspezifischen Datenbankprotokoll aufsetzt. ODBC agiert in diesem Fall wie eine Vermittlungsschicht zwischen dem JDBC-Treiber und den herstellerspezifischen Client Bibliotheken.

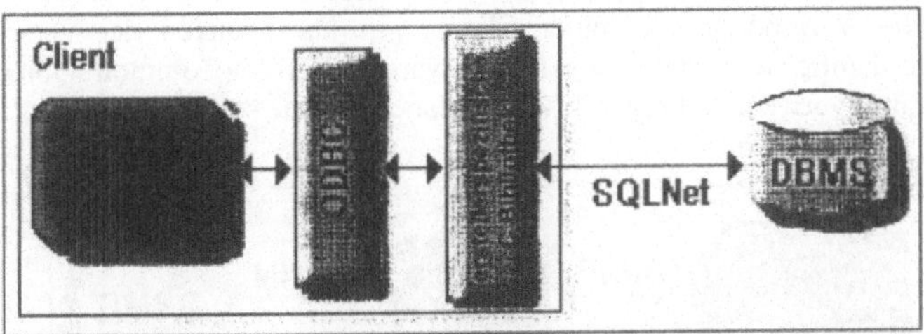

Abbildung 2: JDBC-Zugriff via JDBC-ODBC Brücke auf das DBMS

Die Verwendung dieses JDBC-Treiber Typs setzt eine bestehende ODBC-Installation auf dem Client voraus.

2.1.2 Native-API partly-Java Driver (Type 2)

Eine weiterer Art der JDBC-Treiber Implementierung ist ein sog. Zwei-Schicht-Treiber, der die Client-Datenbankverbindung über vom Datenbankanbieter zur Verfügung gestellte Bibliotheken realisiert. JavaSoft bezeichnet diese Treiber als *"native-API partly-Java drivers"*. Diese JDBC-Treiber wandeln JDBC-Aufrufe in herstellerspezifische Client-API Aufrufe um. Bei Verwendung einer Oracle Datenbank verwenden diese Typ 2 JDBC-Treiber Java Native Methoden, um die C-Schnittstellen der OCI (Oracle Call Interface) Bibliothek aufzurufen.

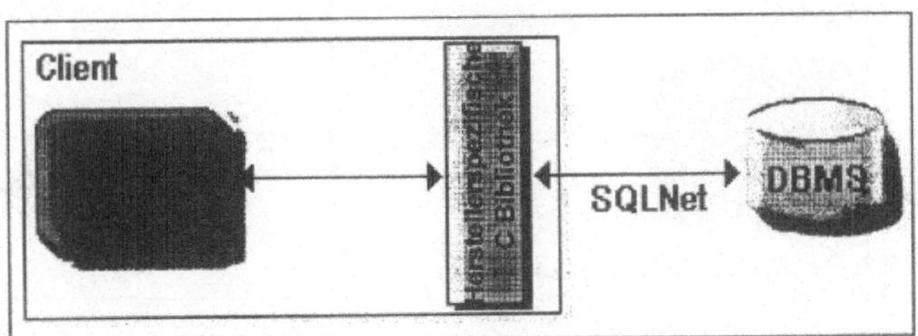

Abbildung 3: JDBC-Zugriff über herstellerspezifische Client Bibliotheken

Vergleichbar wie bei den Typ 1 JDBC-Treibern setzen auch die Typ 2 JDBC-Treiber auf den Clients die Installation herstellerspezifischer Client Bibliotheken voraus wie z.B. SQL*Net bei Oracle. Häufig werden diese Art JDBC-Treiber deshalb auch als "Native"-JDBC Treiber bezeichnet.

2.1.3 JDBC-Net all Java driver (Type 3)

Diese Art JDBC-Treiber wurde für mehrschichtige (Multi-Tier) Umgebungen konzipiert und nicht für direkte Client-Datenbank Kommunikation. Entwickelt und angeboten werden diese Treiber in den meisten Fällen von Middleware-Herstellern, die über Anwendungsserver die Anbindung von Datenbanksystemen unterschiedlicher Hersteller ermöglichen. Die JDBC-Aufrufe werden von diesem JDBC-Treiber in ein DBMS-unabhängiges Netzprotokoll übersetzt, welches dann vom Anwendungsserver in .ein gewünschtes DBMS-Protokoll umgewandelt wird. Diese Anwendungsserver Middleware ermöglicht so die Verbindung all seiner Clients mit Datenbanken der verschiedensten Hersteller. Sehr häufig wird dabei als Middleware CORBA (Common Object Request Broker Architecture) verwendet. Folgende Abbildung zeigt eine solche Umgebung.

Abbildung 4: JDBC-Zugriff über Anwendungsserver

2.1.4 Native-protocoll all-Java driver (Type 4)

Ein vierter Typ von JDBC-Treibern sind die vollständig in Java geschriebenen Treiber die auf ein herstellerspezifisches Protokoll aufsetzen. Diese rufen keine herstellerspezifischen Client-Bibliotheken auf sondern kommunizieren direkt mit dem Datenbanksystem unter Verwendung dessen proprietären Protokolls, das in diesem JDBC-Treiber in Java implementiert ist.

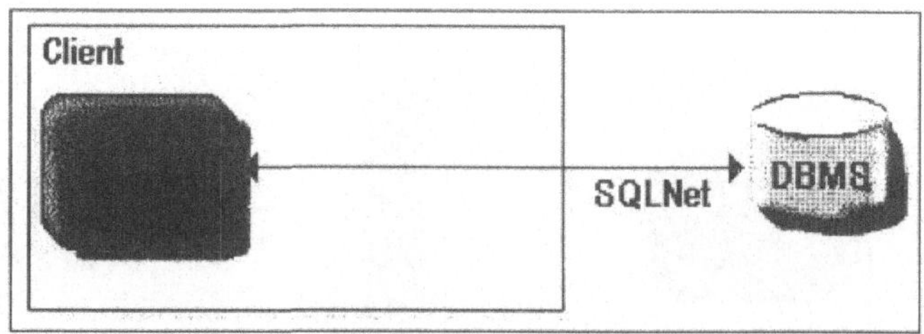

Abbildung 5: Vollständig in Java geschriebener JDBC-Treiber

Bei Oracle wird dieser Treiber als sog. "Thin JDBC Treiber" bezeichnet. Er verwendet Java Sockets um direkt mit dem Listener-Prozeß der Datenbank zu kommunizieren. Er stellt dabei eine eigene Implementierung einer TCP/IP-Version von Oracles SQL*Net zur Verfügung.

2.2 Embedded SQL (SQLJ)

JDBC ist eine generische SQL-Datenbank Schnittstelle (Low-Level-API), die es ermöglicht SQL-Befehlen an eine Datenbank zu schicken. Anders ausgedrückt ist JDBC eine

Schnittstelle, die zur Programmlaufzeit Java-Strings als SQL-Befehle an eine Datenbank übermittelt. Aus diesem Grund spricht man in diesem Zusammenhang von dynamischem SQL. Dynamisches SQL ist sehr flexibel da die SQL-Befehle an die Datenbank erst zur Laufzeit des Programms zusammengestellt werden. Ein Nachteil von dynamischem SQL ist, daß bis zur Ausführung dieses an die Datenbank als SQL-Befehl übermittelten Strings keinerlei Analyse oder Überprüfung des aus diesem String resultierenden SQL-Befehls stattfindet.

Das Gegenstück zu dynamischem SQL ist statisches SQL. Viele Anwendungen benötigen die oben aufgeführte Flexibilität des dynamischen SQL nicht oder nur zu einem geringen Teil, da die in ihnen verwendeten SQL-Befehle von vornherein feststehen. Für diesen Zweck wurde SQLJ, eine gemeinsamen Entwicklung von IBM, Oracle, Sybase und Tandem, entwickelt. Das Ziel von SQLJ ist die Definition eines ANSI/ISO konformen "Embedded SQL" für Java.

SQLJ stellt dafür Java Erweiterungen zur Verfügung, die es ermöglichen, statische SQL-Befehle in Java Programmen einzubetten. Der vollständig in Java geschriebene SQLJ Precompiler ist ein Werkzeug, das diese in Java eingebetteten SQL-Befehle in Standard JDBC-Aufrufe übersetzt. Die Ausgabe des SQLJ Precompilers ist ein generiertes Java-Programm mit JDBC-Befehlen, das von jedem Java-Compiler übersetzt werden kann. Folgende Abbildung verdeutlich den Übersetzungsvorgang von SQLJ.

Abbildung 6: Übersetzungsvorgang eines SQLJ-Programms

Wesentlicher Vorteil bei der Verwendung von statischem SQL für die Programmentwicklung ist die Möglichkeit verschiedene Fehlerquellen bereits während des Precompilerlaufs zu lokalisieren. Dazu gehört die Überprüfung der SQL-Syntax, die Typüberprüfung zwischen den Java und SQL Datentypen und eine Schema-Überprüfung. So wird sichergestellt, daß der SQL-Befehl unter dem Datenbankschema gültig ist, unter dem das SQLJ-Programm laufen soll. Während des Precompilerlaufs muß deshalb eine Verbindung zur Datenbank bestehen.

Ein weiterer wesentlicher Vorteil bei Verwendung von SQLJ ist die direkte Einbindung von Programmvariablen in den SQL-Befehl. Folgender kurzer Programmausschnitt verdeutlicht dies:

Dynamisches SQL (JDBC):
Folgender Programmausschnitt werden die Typen *PreparedStatement* und *ResultSet*, und ihre Methoden (*prepareStatement, setFloat, setInt* und *executeQuery*) von JDBC definiert. Die Variable *recs* enthält das Verbindungsobjekt vom JDBC-Typ *java.sql.Connection*:

```
java.sql.PreparedStatement ps =
      recs.prepareStatement (
      "SELECT STUDENT, SCORE "
    + " FROM GRADE_REPORTS "
    + " WHERE SCORE >= ? "
    + "      AND ATTENDED >= ? "
    + "      AND DEMERITS <= ? "
    + " ORDER BY SCORE DESCENDING");
ps.setFloat(1, limit);
ps.setInt(2, days);
ps.setInt(3, offences);
java.sql.ResultSet honor = ps.executeQuery();
while (honor.next()) {
    System.out.println(honor.getString(1)+" has grade "
            +honor.getFloat(2));
}
```

SQLJ:

Folgender Programmausschnitt ist das äquivalente SQLJ-Programm. Im Gegensatz zu obigem, mit dynamischem SQL entwickelten Beispiel, definiert es ein Typ für die vom SQL-Befehl zurückgegebene Ergebnismenge. Die Java-Variablen werden mit ihrem Variablennamen in den SQL-Befehl eingebunden und nicht über ihre Position.

```
#sql iterator Honors (String name, float grade);
Honors honor;
#sql (recs) honor =
        { SELECT STUDENT AS "name", SCORE AS "grade"
            FROM GRADE_REPORTS
      WHERE SCORE >= :limit
            AND ATTENDED >= :days
            AND DEMERITS <= :offences
      ORDER BY SCORE DESCENDING};
while (honor.next()) {
   System.out.println(honor.name()+" has grade "+honor.grade());
}
```

SQLJ bietet eine große Anzahl Vorteile gegenüber JDBC für statisches SQL:

- SQLJ Quellcode ist vom Umfang her geringer als ein äquivalentes JDBC-Programm.
- Beim Precompilerlauf besteht eine Datenbankverbindung, was die Typ- und Syntaxüberprüfung der SQL-Befehle ermöglicht. JDBC, das vollständig dynamisch ist, bietet diese Überprüfungsmöglichkeiten nicht.
- In SQLJ-Programmen können die Programmvariablen direkt in den SQL-Befehl eingebettet werden. JDBC hingegen benötigt ein separaten Call-Befehl für jede Programmvariable und spezifiziert die Bindung über die Positionsnummer.
- SQLJ stellt einfachere Regeln für den Aufruf von Stored Procedures und Functions zur Verfügung.

Insgesamt gesehen stellt SQLJ eine wesentliche Vereinfachung bei der Entwicklung von Datenbankanwendungen in Java dar.

2.3 Common Object Request Broker Architecture (CORBA)

Wie bereits in der Einleitung erwähnt wird in der Softwareentwicklung mehr und mehr ein neues Architekturmodell, das Drei-Schichten-Modell verwendet. Dabei werden die Bereiche Daten, Unternehmenslogik und Benutzerschnittstelle voneinander getrennt. Dieses Konzept ist bei Entwicklern auch als "Model-View-Controller" Entwurfsmuster bekannt [5]. Durch die Trennung von Daten, Unternehmenslogik und Benutzerschnittstelle lassen sich sehr schnell Nutzen erzielen, die weit über die des zweischichtigen Client/Server-Ansatzes hinausgehen, die des verteilten Computings. Die CORBA-Technologie der OMG (Object Management Group) ist für dieses Modell geradezu prädestiniert [8].

Vorteilhaft sind in diesem Zusammenhang die Gemeinsamkeiten der Objektmodelle von CORBA und Java, beispielsweise durch das Schnittstellen-Konzept. Dadurch ist es möglich, das Objektmodell von CORBA fast nahtlos in Java umzusetzen, was die Softwareentwicklung verteilter Anwendungen wesentlich erleichtert. CORBA ermöglicht nicht nur eine transparente Verteilung von Java-Objekten in einem Netz, sondern auch den Zugriff auf beliebige andere CORBA-Objekte, egal in welcher Programmiersprache diese vorliegen. Dadurch lassen sich auch sehr einfach bereits vorhandene Anwendungen einbinden. Folgende Abbildung zeigt die Architektur von CORBA.

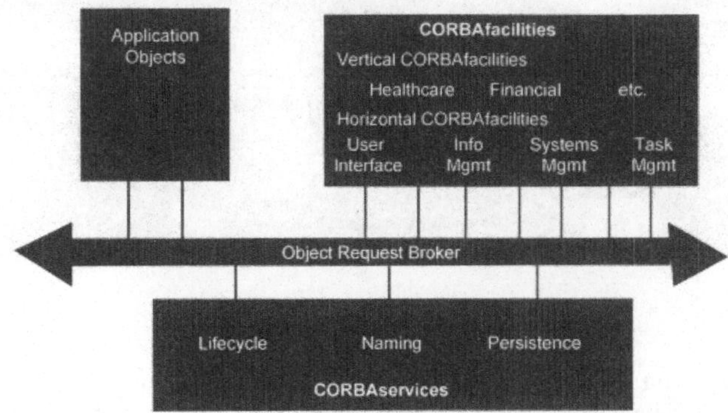

Abbildung 7: Die Object Management Architektur der OMG

Zentraler Bestandteil von CORBA ist der Object Request Broker (ORB). Dessen primäre Aufgabe besteht darin die Methodenaufrufe von einer Client-Anwendung zu einem entfernten

Objekt zu übermitteln und evtl. auftretende Fehler und die Ergebnisse an den Client zurück zu melden.

Vorteile beim Einsatz von CORBA ist neben einer freien Wahl der Programmiersprache für die verteilten Objekte auch die Plattformunabhängigkeit. CORBA-Implementierungen laufen unter praktisch allen Unix-Varianten, OS/2, OS/400, Mac-OS, VME, MVS, VMS sowie MS-DOS, 16-Bit und 32-Bit-Windows.

3. Architekturmodelle

3.1 2-Tier vs. 3-Tier

2-Tier Modelle kommen aus der klassischen Client/Server Datenbankprogrammierung. Dort wird aus Client-Anwendungen, die in 3/4GL Sprachen entwickelt sind, über unterschiedliche Netzprotokolle auf relationale Datenbanken zugegriffen. Hierbei kommen allgemeine Schnittstellen wie ODBC oder proprietäre DB-API's (z.B. SQL*Net von Oracle) zum Einsatz.

Multi-Tier-Modelle werden nun eingesetzt um Client/Server Applikationen in funktionale Einheiten aufzusplitten. Diese Einheiten werden dem Client und ein oder mehreren Servern zugeteilt. Der wohl typischste Vertreter dieser Architektur ist das 3-Tier Modell, in der auf Client-Seite das reine User-Interface abläuft, in der Middle-Tier die Business-Logik realisiert wird und auf dem dritten Layer die Datenhaltung liegt. Die Kommunikation zwischen den einzelnen Schichten wird über eine Middleware (z.B. CORBA) implementiert.

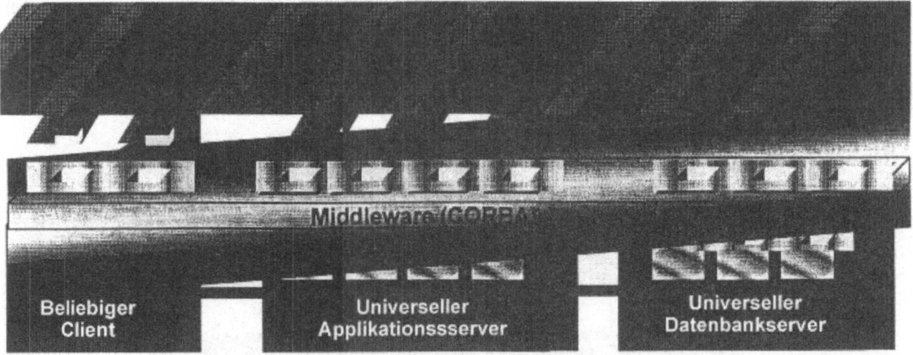

Abbildung 8: 3-Tier Architektur

Bei Verwendung von CORBA ist es nun möglich auf den einzelnen Schichten Objekte (*Komponenten*) beliebig auszutauschen. So ist z.B. eine Client Komponente, die für eine Windows-Anwendung geschrieben ist, ohne Rekompilierung der kompletten Anwendung auf eine Java-basierte Internet-Anwendung umstellbar. Reale Implementierungen der vorgestellten 3-Tier Architektur sind Oracle's Network Computing Architecture [9] oder Netscape ONE [6].

In Client/Server Anwendungen muß die Business-Logik entweder im User-Interface des Clients oder als "Stored procedures" in der Datenbank implementiert werden. Beide Alternativen, die im Regelfall sowohl als auch eingesetzt werden, haben den Nachteil, daß sie zu einer Überladung der Schichten führen. "Fat Clients" sind die Folge, die neben einem enormen Ressourcenverbrauch auch sehr viel Arbeit auf Administrationsseite verbuchen. Für Intra/Internet-Applikationen sind Client-seitige Installationen überhaupt nicht oder nur mit extremem Aufwand zu realisieren.

Eine Aufteilung funktionaler Gesichtspunkte auf unterschiedliche Schichten bietet enorme Vorteile in Bezug auf Skalierbarkeit, Performanz und Funktionalitätserweiterungen. Zunächst ist eine Reduzierung der Netzwerkaktivitäten durch die *Kanalisierung* von Serveranfragen möglich. In der Middle-Tier wird ein Pool von Anfragen gesammelt und dann über eine Verbindung zum Datenbankserver abgewickelt. Die Tatsache, daß ein Server in der Middle-Tier im Regelfall eine weitaus performantere Maschine darstellt als ein normaler Arbeitsplatzrechner, führt zu Geschwindigkeitsvorteilen in der Abarbeitung komplexer Geschäftsprozesse. Erweiterungen der Funktionalität ist durch das bereits beschriebene Komponentenmodell realisierbar. Zusätzlich bieten Multi-Tier Anwendungen bei Einsatz von *TP-Monitoren* [1] eine große Vielfalt an Messaging-Techniken an, die zu einem hohen Maß an Flexiblität in Bezug auf *Load-Balancing* führen.

3.2 Thin Client 2-Tier Architektur

Bei einer 2-Tier Architektur handelt es sich um eine klassische Client-Server Anwendung, die über eine gängige Netzverbindung Daten zwischen der Applikation (Client) und der Datenbank (Server) austauscht. Beim hier vorgestellten Modell wird die bereits beschriebene Java Database Connectivity (JDBC) benutzt. Dies bedeutet, daß die Anwendung entweder direkt JDBC-Calls oder embedded SQL (SQLJ) benutzen kann. Abbildung 9 illustriert das Szenario:

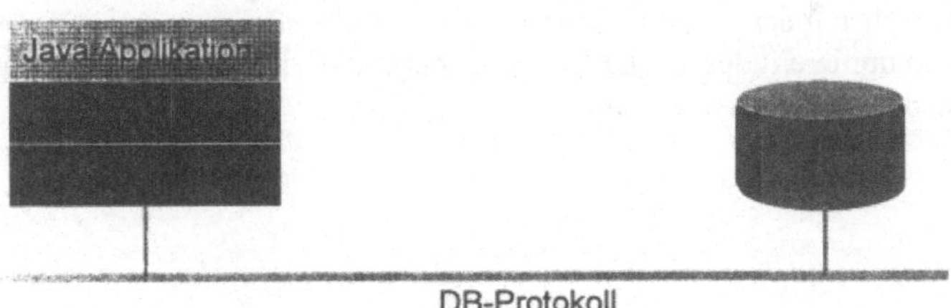

Abbildung 9: 2-Tier Architektur basierend auf JDBC

Eine Java-Applikation greift bei dieser Architektur über JDBC-Calls direkt auf die Datenbank zu. Dabei können alle gängigen Typen von JDBC-Treibern benutzt werden. In Abbildung 9 wird ein DB-spezifisches Netzwerkprotokoll (z.B. SQL*Net von Oracle) eingesetzt, welches JDBC-Treiber vom Typ 2 (*native-API partly-Java driver*) und Typ 4 (*native-protocol all-Java driver*) zuläßt.

Gegen die generelle Verwendung von Typ 1 Treiber (*JDBC-ODBC bridge*) sprechen u.a. folgende Gründe:

- **Overhead**: JDBC-API Calls werden zur JDBC-ODBC Bridge übertragen und dort in die ODBC API in Form der Programmiersprache C umgewandelt. Anschließend findet ein erneutes Mapping von ODBC in die API der angesprochenen Datenbank statt. Der gesamte Umwandlungsprozess ist ressourcenfressend und nicht performant.
- **Funktionalitätsverlust**: ODBC Treiber arbeiten auf der kleinsten gemeinsamen Funktionalitätsebene verfügbarer Datenbanken. Anbieterspezifische Erweiterungen hinsichtlich Funktionalität, Skalierbarkeit und Performanz sind im Regelfall durch ODBC nicht zu realisieren.

Auf eine Betrachtung des Typ 3 Treibers (*net-protocol all-Java driver*), der auf einer dritten Schicht zur Kommunikation zwischen Client und Server aufbaut, wurde zugunsten CORBA-basierter 3-Tier Architekturen an dieser Stelle verzichtet.

Für weiterführende Vergleiche werden wir uns auf eine Implementierung aufbauend auf einem Treiber vom Typ 4 beziehen. Ziel ist es einen möglichst leicht administrierbaren Thin-Client zu untersuchen, der sowohl in einer Intranet als auch in einer Extranet-Umgebung einsetzbar ist. Typ 4 Treiber benötigen auf Client-Seite keinerlei Installationsaufwand und werden bei einer Größe von ca. 150 KB[2] bei Start einer Applikation auf den Client geladen. Aus diesem Grund eignen sie sich neben dem Einsatz in einer reinen Java-Applikationen auch für Web-basierte Anwendungen im Intra/Extranet.

Die vorgestellte 2-Tier Architektur hat den Nachteil, daß jeder Anwender eine direkte Verbindung zur Datenbank benötigt. Dies hat neben extremem Netzwerkverkehr auch negative Auswirkungen auf die Performanz der Datenbank. Desweiteren zwingen Änderungen am Datenmodell oder an der Applikationslogik zur Reprogrammierung der kompletten Anwendung.

Eine Erleichterung der Programmierung auf Client-Seite ist sicherlich durch Einsatz von SQLJ zu erreichen. Desweiteren kann durch SQLJ eine Flexibilität in Bezug auf Ausnutzung herstellerspezifischer Features gewonnen werden. Dies geht jedoch nur gering auf Kosten der Portabilität, denn SQLJ-Precompiler können aus gleichem Java-Code jeweils herstellerspezifischen oder standardisierten JDBC-Code erzeugen. In anderen Worten – einfaches Rekompilieren des SQLJ-Codes genügt um die Anwendung gegen jegliche relationale Datenbank laufen zu lassen.

---

[2] Diese Größe bezieht sich auf den komprimierten Typ 4 JDBC Treiber der Oracle Corporation.

3.3 Thin-Client 3-Tier IIOP/DB-Protokoll Architektur

Bei folgender Architektur wird ein 3-Schichtenmodell eingesetzt. Die einzelnen Schichten sind über unterschiedliche Netzprotokolle miteinander verbunden.

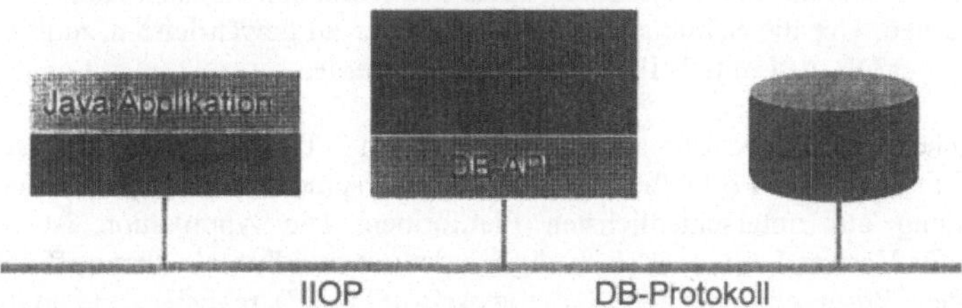

Abbildung 10: 3-Tier IIOP/DB-Protokoll Architektur

Beim vorgestellten Modell wird zwischen dem Client und der Middle-Tier das CORBA-Netzwerkprotokoll Internet-Inter-ORB Protokoll (*IIOP*) [8], und zwischen Middle-Tier und Server ein proprietäres Datenbank-Netzwerkprotokoll (z.B. SQL*Net von Oracle) eingesetzt. Zum effektiven Zugriff auf die Datenbank wird ein C++-Adapter benutzt, der über eine Datenbank-API (z.B. das Oracle Call Interface) direkt die Datenbank adressiert. Der C++-Adapter ist in der Interface Definition Language (*IDL*) spezifiziert und kann aus diesem Grund von beliebigen CORBA-Objekten angesprochen werden. Auf Client-Seite werden nun Klassen innerhalb der Java-Applikation zum Datenzugriff ebenfalls in IDL spezifiziert.

Bei Programmierung des Clients sind keinerlei Datenbankkenntnisse notwendig. Der transparente Zugriff auf relationale Daten wird durch die Middle-Tier übernommen. In Bezug auf Flexibilität wäre ein Austausch der Datenzugriffsschicht z.B. auf flache ASCII-Dateien oder Host-Daten ohne Umstellung der eigentlichen Applikation problemlos möglich. Dies bedeutet weiterhin, daß keine JDBC-Aufrufe innerhalb des Clients mehr notwendig sind, sondern Daten über reine Methodenaufrufe adressiert werden.

3.4 Thin-Client 3-Tier CORBA Architektur

Ähnlich dem zuvor beschriebenen Modell werden in dieser Architektur drei Schichten verwendet:

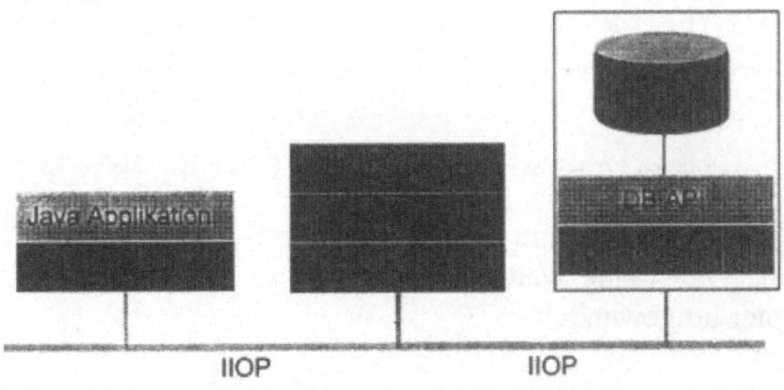

Abbildung 11: 3-Tier CORBA Architektur

Die komplette Kommunikation zwischen den einzelnen Tiers findet über IIOP statt. Dazu werden die Schnittstellen über IDL spezifiziert. Im Vergleich zur vorherigen 3-Tier Architektur muß am Client, der seine Anfragen an das Datenhaltungssystem über

Methodenaufrufe formuliert, nichts verändert werden. Die Middle-Tier wird durch eine auf Java basierende Schicht komplett ersetzt. Eine Verlagerung der Datenbank-API von der Middle-Tier auf die Server-Tier wird vorgenommen, wobei kaum Änderungen am Source-Code notwendig sind. Da die DB-API nun auf der gleichen Maschine wie die eigentliche Datenbank läuft, kann auf die Verwendung eines datenbank-spezifischen Netzwerkprotokolls verzichtet werden. Um die Schnittstelle zur Middle-Tier zu gewährleisten, muß jedoch eine Beschreibung der DB-API mittels IDL vorgenommen werden.

Ein wesentlicher Unterschied zum anderen vorgestellten 3-Tier Modell ist die fast komplette Realisierung in Java. Die Portabilität der Programmiersprache Java führt zu Verfügbarkeit der Anwendung auf unterschiedlichsten Plattformen. Die Applikation ist demzufolge prädestiniert für Verwendung im Web. Lediglich der unterste Datenbankzugriff wird in einer datenbanknahen Programmiersprache (vorzugsweise C/C++[3]) realisiert und muß bei einer Portierung neu kompiliert werden. Desweiteren findet in diesem Modell eine exaktere und restriktivere Trennung von Business-Logik und DB-Zugriff statt. Im Gegensatz zur vorherigen 3-Tier Architektur werden alle Prozesse, die direkt mit der Datenbank kommunizieren (einfügen, löschen, modifizieren von Datensätzen) direkt auf der Server-Tier abgewickelt. Nur komplexe, funktionale Prozesse werden auf der Middle-Tier ausgeführt. Bei simplen Abfrage-Routinen kann die Middle-Tier zum Mapping zwischen Datenbank- und Endbenutzersicht verwendet werden.

4. Performanz Gesichtspunkte

Innerhalb einer Performanz-Analyse wollen wir unterschiedliche Abarbeitungszeiten für 2 bzw. 3-Tier Architekturen betrachten. Dabei beschränken wir uns auf das bereits beschriebene 2-Tier Modell und eine abgewandelte 3-Tier Architektur. Um einen objektiven Performanzvergleich ziehen zu können, wurde der Client der 3-Tier Applikation keiner logischen Veränderung unterzogen. Dies bedeutet, daß auch etwaige JDBC-Aufrufe belassen und in IDL deklarierte Klassen gekapselt wurden. Eine 3-schichtige vergleichbare Architektur hat nun folgenden Aufbau:

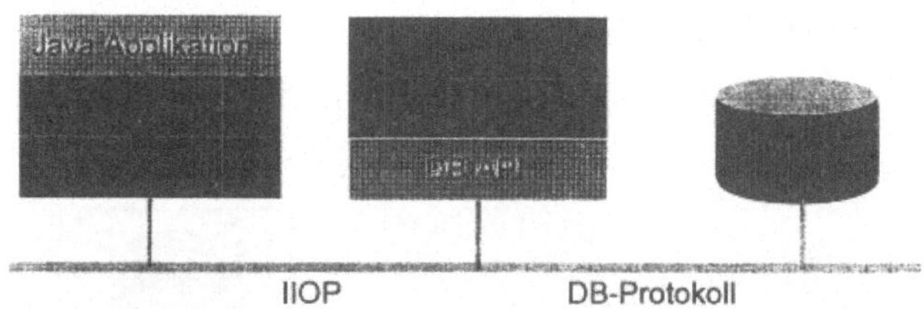

Abbildung 12: 3-Tier Architektur mit JDBC-Calls in der Client-Tier

Im Vergleich zur 2-Tier Architektur werden nun Anfragen nicht mehr über den JDBC-Treiber direkt an die Datenbank weitergeleitet, sondern in datenbankspezifische Methoden über den C++-Adapter umgewandelt.

Basis dieser Umwandlung ist die Verwendung von SQL-Calls, die 1:1 abgebildet werden:

---

[3] Zum jetzigen Zeitpunkt hat kein großer Datenbankhersteller eine low-level DB-API in Java zur Verfügung gestellt. Standardzugriff auf relationale Datenbanken soll nur über JDBC erfolgen. Eine Realisierung kann somit nur über JDBC auf Serverseite stattfinden.

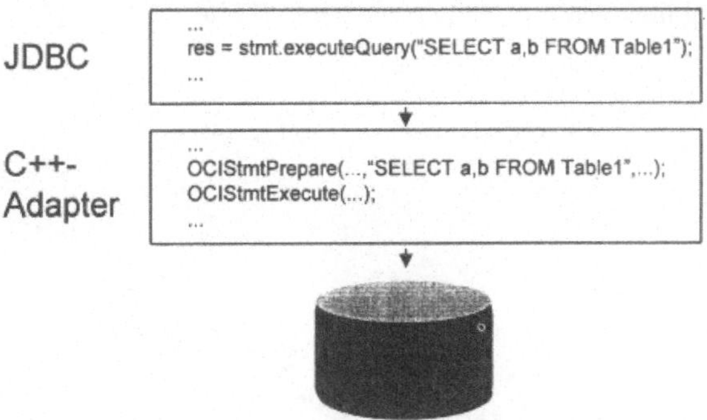

Abbildung 13: Umwandlung JDBC in DB-API

Innerhalb des Benchmarks wird nur jeweils lesend auf die Datenbank zugegriffen. Jeder dieser lesenden Zugriffe liefert eine vordefinierte Anzahl Datensätze: 50, 100, 500, 1000, 2000 und 10000 Zeilen.

Zusätzlich zum C++-Adapter werden unterschiedliche Caching-Strategien in der Middle-Tier verwendet. Bei *synchronem Caching* werden zeitliche Verzögerungen zwischen den Tiers bemerkt, die linear proportional zur Anzahl der eingesetzten Tiers sind [3]. Daten zwischen zwei Tiers werden in Blöcken übertragen, die eine vordefinierte Anzahl von Datensätzen (*row prefetch)* enthalten. Zeitliche Verzögerungen kommen nun dadurch zustande, daß der Client nach Abschicken seiner Anfrage darauf warten muß, daß zurückgelieferte Datensätze in einen Block zusammengefaßt werden. Bei synchroner Verarbeitung kann der nächste Block erst angefordert werden, wenn der empfangene Block abgearbeitet wurde. Um dieses Zeitintervall zu überbrücken können nun Threads eingesetzt werden. Sobald der Client den ersten Datensatz im Datenblock bearbeitet, wird ein Thread kreiert der bereits die Anfrage nach dem nächsten Block beinhaltet. Während nun der Child-Thread auf Datensätze wartet, können die im Parent-Thread empfangenen Datensätze der Client-Tier zur Abarbeitung zur Verfügung gestellt werden (*asynchrones Caching*). Diese Multithreading-Strategie wird in der Middle-Tier implementiert und liefert erstaunliche Performanzgewinne.

Folgende Abbildung illustriert die relativ gemessenen Werte:

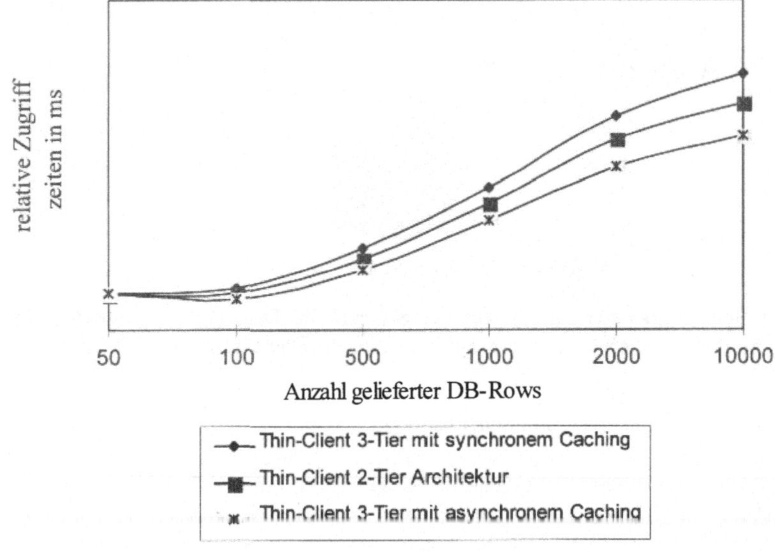

Abbildung 14: Performanzvergleich unterschiedlicher Architekturen[4]

Wie vermutet werden bei kleinen Datenmengen ähnliche Werte erreicht. Bei steigender Anzahl empfangener Datensätze setzt sich die traditionelle 2-Tier Client/Server Architektur gegenüber synchronem 3-Tier mit ca. 14% besserer Performanz durch. Dies illustriert die genannte zeitliche Verzögerung pro Schicht und ist konform mit Messungen in [7]. Vergleicht man nun Client/Server mit optimiertem 3-Tier, d.h. asynchronem Caching, wird eine Performanzsteigerung um durchschnittlich weitere 15% festgestellt.

Es bleibt zu bemerken, daß in diesem Benchmark Single-User Belastung getestet wurde, was nur die Grundperformanz der einzelnen Technologien darstellt. Wir erwarten weitere Performanzgewinne, sowohl bei asynchronem als auch bei synchronem Caching, gegenüber Client/Server durch den Effekt der Kanalisierung von Datenbankverbindungen bei 3-Tier Architekturen.

5. Fazit

Im vorliegenden Paper haben wir einen Überblick über Basis-Technologien zum Zugriff auf relationale Datenbanken von der Programmiersprache Java aus beleuchtet. Wir zeigten unterschiedliche Multi-Tier-Architekturen auf und diskutierten Vor- und Nachteile in Bezug auf Erweiterbarkeit des Codes, Flexibilität und Performanz der Anwendung. Im Detail wurde eine 2-Tier und 3-Tier Architektur ohne Abänderung der Client-Seite auf Performanz getestet. Es konnte gezeigt werden, daß 3-Tier Anwendungen durch Optimierung der Middle-

---

[4] Die Werte stammen aus Oracle internen Performanztests [3].

Tier schnelleren Datenbankzugriff gewährleisten als traditionelle Client/Server-Anwendungen.

Grundsätzlich liegen die Vorteile von Multi-Tier Architekturen klar auf der Hand:

- **Flexibilität/Funktionalität:** Bei Einsatz einer Middleware wie z.B. CORBA ist es möglich verschiedene Komponenten einzusetzen. Interoperabilität ist hierbei durch die Middleware gewährleistet. Zusätzlich kann auf unterschiedliche Laufzeitanforderungen flexibel durch Load-Balancing oder diverse Messaging-Techniken reagiert werden.
- **Skalierbarkeit:** Middleware bietet extreme Skalierbarkeit durch Kanalisierung von DB-Requests als auch durch Einsatz mehrerer Applikations/DB-Server auf die transparent zugegriffen wird.
- **Ausfallsicherheit:** Durch Verwendung mehrerer identischer Server kann die Ausfallsicherheit von "Mission-Critical" Anwendungen erheblich verbessert werden.
- **Performanz:** Hingegen der Grundregel, daß das Zwischenschalten mehrerer Tiers zur Geschwindigkeitsreduzierungen führt, konnte gezeigt werden, daß Optimierungstechniken einen deutlichen Gewinn an Performanz von Multi-Tier Applikationen gegenüber Client/Server-Applikationen bringen.

Abschließend kann festgestellt werden, daß Java die Sprache der Wahl für Inter/Intranet-basierende Anwendungen darstellt. Neben portablen Anwendungen und Thin-Client Unterstützung sind auch stabile, performante und standardisierte Möglichkeiten vorhanden aus Java auf relationale Datenbanken zuzugreifen. Inzwischen bieten alle großen Datenbankhersteller eine JDBC-Schnittstelle an. Welche Rolle Java inzwischen spielt hat u.a. eine Ankündigung von Oracle gezeigt, die besagt, daß eine Java Virtual Machine direkt im Adressraum der Oracle8 Datenbank in einem der nächsten Releases laufen wird. Damit wird es möglich sein Java-Methoden als "Stored Procedures" in der Datenbank zu verwenden.

Literatur

[1] B. Coleman. *Destination: The Supernet.* http://www.beasys.com/action/unican.htm. BEA Systems, 1997.

[2] D. Curtis. *Java, RMI and CORBA.* http://www.omg.org/news/wpjava.htm. Object Management Group Whitepaper, 1997.

[3] A. Ezzat, J. Fernandez, K. Ketefian, J. Li. *Multi-Tier Applications Performance Analysis.* Oracle internal Whitepaper, Oracle Corporation, Redwood Shores, USA. 1997.

[4] A. Ezzat, J. Fernandez, K. Ketefian, J. Li. *Thin-Client, 3-Tier JDBC Architecture.* Oracle Whitepaper, Oracle Corporation, Redwood Shores, USA. 1997.

[5] E. Gamma, R. Helm, R. Johnson, J. Vlissides. *Design Patterns - Elements of Reusable Object-Oriented Software,* Addison-Wesley, 1995

[6] The Networked Enterprise. Netscape Communications Corporation, Mountain View, USA, 1997.

[7] R. Orfali, D. Harkey. *Client/Server Programming With Java and Corba.* John Wiley & Sons, USA, 1997.

[8] *CORBA.Architecture and Specification.* http://www.omg.org/corba/corbiiop.htm OMG, Framingham, USA, 1997.

[9] Network Computing Architecture. Oracle Deutschland GmbH, München, 1996.

[10] *Bringing Java to the Enterprise.* Oracle Corporation, Redwood Shores, USA, 1997.

[11] *Oracle's JDBC Driver – Accessing the Oracle RDBMS from Java.* Oracle Corporation, Redwood Shores, USA, 1997. http://www.oracle.com/products/free_software/

[12] *SQLJ: Embedded SQL for Java.* Oracle Corporation, Redwood Shores, USA, 1997

[13] *SQLJ User Guide and Reference.* Oracle Corporation, Redwood Shores, USA, 1997

[14] P. Patel, K. Moss. *Java Database Programming with JDBC.* Coriolis Book Group, USA, 1997.

[15] *JDBC Guide: Getting Started.* Sun Microsystems, Mountain View, USA, 1995, 1996.

[16] *JDBC Drivers.* http://java.sun.com/products/jdbc/jdbc.drivers.html. Sun Microsystems, Mountain View, USA, 1997.

[17] *The JDBC Database Access API.* http://java.sun.com/products/jdbc. Sun Microsystems, Mountain View, USA, 1997.

[18] Joel Scotkin. *Java - Das Drei-Schichten-Konzept.* SIGS Conferences, Java Spektrum, Juli/August 1996, S. 53ff.

[19] A. Voge, K. Duddy, A. Vogel. *Java Programming With Corba.* John Wiley & Sons, USA, 1997.

[20] *VisiBroker for Java/C++.* http://www.visigenic.com/info/prod.html. Visigenic, San Mateo, USA, 1997.

[21] *Choosing a JDBC Driver.* http://www.weblogic.com/whitepapers/jdbc.html. WebLogic Inc., San Francisco, USA, 1997.

[22] *Developers Guide for jdbcKona/Oracle.* http://www.weblogic.com/classdocs20/API_joci.html. WebLogic Inc., San Francisco, USA, 1997.

[23] *JDBC Driver Diagram.* http://javanese.yoyoweb.com/JDBC/DriverTypes.html. YoYo Publications, Los Angeles, USA, 1997.

Adressen der Autoren

Horst Wend
Oracle Deutschland GmbH
Forstenrieder Allee 61
D-81476 München
e-mail: hwend@de.oracle.com
Tel: +49 (0)89 75981-167
Fax: +49 (0)89 75981-107

Timo Salzsieder
Oracle Deutschland GmbH
Forstenrieder Allee 61
D-81476 München
e-mail: tsalzsie@de.oracle.com
Tel: +49 (0)89 75981-274
Fax: +49 (0)89 75981-107

Web-basiertes, kooperatives Lernen auf Basis dynamischer Gruppenbildung

Karin Schmidt, Johannes Bumiller, Peter Manhart

Daimler-Benz AG, Forschung Informationstechnik

Zusammenfassung

Hohe Innovationszyklen, wechselnde Rahmenbedingungen und laufend neue Anforderungen, erfordern ständiges Lernen, von jedem Einzelnen, Gruppen und auch von Organisationen. Daraus resultiert ein wachsender Bedarf an geeigneten lernunterstützenden Systemen. Dieser Beitrag beschreibt ein Web-basiertes System zur Unterstützung kooperativen Lernens in einer anpaßbaren, offenen Umgebung. Im Gegensatz zu vielen vorhandenen, traditionellen Systemen konzentriert sich dieser Ansatz auf gruppenorientiertes Lernen, im besonderen auf die Aspekte (1) Wahrnehmung anderer Lernenden, (2) dynamische Gruppenbildung, in Abhängigkeit von den Interessen und Wünschen des Lernenden und (3) Unterstützung der Interaktion.

1. Einführung

Das Wissen der Menschheit steigt exponentiell an. Wechselnde und neu entstehende Anforderungen und Rahmenbedingungen in allen Bereichen unseres Lebens (insbesondere durch neue Technologien) erfordern "lebenslanges Lernen". Viel stärker als früher, werden die Phasen Grundausbildung, Spezialisierung und anschließende Umsetzung des erworbenen Wissens im Beruf, unterbrochen durch ständige, ineinander verwobene Zyklen des Wissenserwerbs, dessen Anwendung und auch dessen Verfalls. Individuen, wie auch Organisationen, sind gezwungen, ständig neues Wissen aufzubauen, zu erhalten und auch verfügbar zu machen bzw. weiterzugeben.

Daraus entsteht ein stetig wachsender Bedarf an strukturierter, den Anforderungen des Lernenden angepaßter Information, wie auch an neuen, flexiblen Möglichkeiten zur Unterstützung des Lernens abseits der traditionellen Schulausbildung. Ein wichtiges, aber (auch auf Basis innovativer Technologien) oft vernachlässigtes Element des Lernens ist der soziale Kontext (d.h. Lernen zusammen mit anderen) - wann immer isoliertes Arbeiten nicht sinnvoll oder erwünscht ist. Zur Unterstützung dieses gemeinsamen Lernens muß als Grundanforderung ein gemeinsamer Informationsraum verfügbar sein, der einfach zugänglich und nutzbar ist, die Wahrnehmung anderer ermöglicht und Mittel zur Kommunikation untereinander bereitstellt.

Das World Wide Web ist hardware- und softwareunabhängig, kostengünstig und leicht zugänglich und deshalb auch weit verbreitet. Es unterstützt die Informationsbereitstellung und -verteilung sehr effizient und entwickelt sich immer mehr zu einer Plattform, die auch für komplexere Dienste nutzbar ist. Im Kontext des kooperativen Lernens ist das Web deshalb als verteilte, leicht nutzbare Plattform zur Interaktionsunterstützung sehr gut einsetzbar.

Dieser Beitrag beschreibt zuerst Anforderungen und eine prototypische Implementierung eines Systems zur Unterstützung des kooperativen Lernens (Kapitel 2). Anschließend werden spezielle Mechanismen für den dynamischen Aufbau von Lerngruppen vorgestellt (Kapitel 3). Kapitel 4 zeigt dann verschiedene Möglichkeiten der Gruppeninteraktion auf, während Kapitel 5 die Basisarchitektur des Prototypen kurz beschreibt. Ein kurzer Überblick über verwandte Arbeiten und eine Zusammenfassung schließen den Beitrag ab.

2. Anforderungen und Anwendungs-Szenarium

Lernen in der oben beschriebenen Umgebung, basiert auf einer verteilten Struktur und erfordert die Unterstützung verschiedener, auch lose gekoppelter Interaktionsmechanismen. Der Dienst kooperatives Lernen ist nicht auf das Lernen in statischen Klassen mit festgelegten Rollen (Lehrer, Schüler) und Aufgaben ausgerichtet. Der Schwerpunkt liegt vielmehr auf der Unterstützung gemeinsamen Lernens in einer flexiblen, bezüglich Inhalt und Teilnehmern, sich ändernden Umgebung. Daraus resultieren Fragestellungen, wie z.B.: Wo kann ich relevante Information finden ? Wer sonst interessiert sich für diese Information ? Gibt es jemanden, der mir dazu helfen kann ? Ist jemand am selben Thema, am gemeinsamen Lernen dazu interessiert ? Kann ich jemanden helfen ? Wie lernen andere ?

Drei wesentliche Anforderungen bestimmen Unterstützungsmöglichkeiten für diese Form des kooperativen Lernens: (1) Die Wahrnehmung anderer Lernenden, (2) der flexible Aufbau von verteilten Lerngruppen und (3) bedarfsgerechte Interaktionsmöglichkeiten zwischen den Lernenden.

Die Wahrnehmung anderer Personen ist eine Grundvoraussetzung für Gruppenbildung und Gruppeninteraktion [2]. Ein wesentlicher Aspekt des kooperativen Lernens ist, wer mit wem auf welche Art und Weise in Kontakt tritt. Im beschriebenen Ansatz werden dazu generische Maße, sogenannte *Metriken* definiert, die eine dynamische und flexible Gruppenbildung erlauben. Jeder Benutzer kann sich in verschiedenen Kategorien beschreiben, z.B. vorhandene Kenntnisse, Interessen, Absichten. Diese Informationen können genutzt werden, um Lernende über einen Gruppenbildungsmechanismus zusammenzuführen. Neben der Möglichkeit, Benutzer mit ähnlichen Interessen zu finden, kann durch Gruppenbildung auch ein sozialer Kontext hergestellt und dadurch Interaktion vereinfacht werden – aus einem Individuum wird ein Gruppenmitglied. Verschiedene Verfahren zur Gruppenbildung resultieren in verschiedenen Typen von Lerngruppen. Abhängig von Gruppentypen und Benutzerinteressen sind verschiedene Interaktionsmechanismen notwendig. Die Unterstützung der Gruppen-interaktion orientiert sich dabei an den spezifischen Bedürfnissen einer Lerngruppe.

Zur Überprüfung der Konzepte auf Nutzbarkeit und Leistungsvermögen wurden Prototypen für ein allgemeines Gruppeninteraktionssystem entwickelt. Das folgende Szenario veranschaulicht die Schwerpunkte des web-basierten kooperativen Lernens.

Bildschirm Bereiche

(a) Browser Toolleiste

(b) System Funktionsgruppen

(c) Übersichtsbereich

(d) Chat Bereich

(e) Gruppenbereich

(f) Inhalt

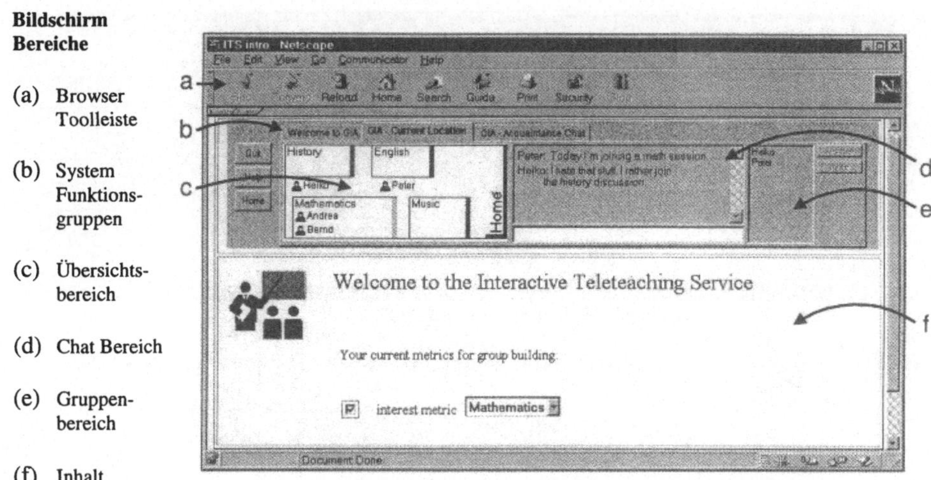

Abb. 1: Der Teleteaching Service für kooperatives Lernen

Sobald die Eingangsseite des Demonstrators geladen wird, teilt sich der Browser in zwei Bereiche auf. Im oberen Bereich läuft der Web Client als Java Applet. Im unteren Teil wird die Homepage des Teleteaching Service dargestellt (Abb. 1, Teil f). Aus der Sicht des Teilnehmers Peter soll die Funktionsweise erklärt werden. Da die Funktion "GIA - Current Location" (Abb. 1, Teil b) aktiviert wurde, werden im Übersichtsbereich (Abb. 1, Teil c) alle für Peter sichtbaren Lernräume, zusammen mit seiner aktuellen Position, angezeigt. Neben dem Teilnehmer Heiko, der sich mit ihm im gleichen Bereich befindet, sind auch alle anderen Teilnehmer in den für ihn sichtbaren Räumen, dargestellt. Peter und Heiko werden zunächst in einer Gruppe zusammengefaßt (Abb. 1, Teil e). Die Gruppierung basiert hier auf einer räumlichen Zuordnung mit der Link-Distanz 1 (Beschreibung folgt in Kapitel 3). Peter entscheidet sich für einen Wechsel in den Mathematikraum. Das System erlaubt ihm beim Eintritt in einen anderen Raum, zwischen vordefinierten Einträgen zu wählen. Diese Einträge sind verfügbar über eine Aufklappleiste. Beispielsweise kann für den Mathematikraum zwischen den Fächern linearer Algebra, Analysis, Stochastik, Numerik usw. entschieden werden. Die Liste der Einträge ist einfach erweiterbar. Vorgesehen ist, daß der Benutzer dabei auch auf beliebige Kombinationen zurückgreifen kann.

Peter entscheidet sich für den Eintrag "Analysis" im Mathematik-Menü und betritt das Mathematik-Klassenzimmer (Abb. 2). Da die angegebenen Benutzerinteressen von Bernd und Andrea zu seiner eigenen passen (zu bestimmen durch die Anwendung der Funktion Int-Match, Metrik der Benutzer-Interessen, Kapitel 3), werden die drei Teilnehmer in einer Gruppe zusammengefaßt. Sie können nun mit Hilfe von Kommunikationswerkzeugen, z.B. dem Chatting Bereich (Abb. 1, Teil d), über dargestellte Lerninhalte diskutieren.

Die Unterstützung im Anwendungsszenario "Telelernen" konzentriert sich auf automatische, dynamische Gruppenbildung, basierend auf den Wünschen und Interessen der Teilnehmer. In den meisten vorhandenen Systemen zur Unterstützung des Telelernens, basieren Gruppen dagegen auf den fixierten Rollen von Lehrer und Schüler.

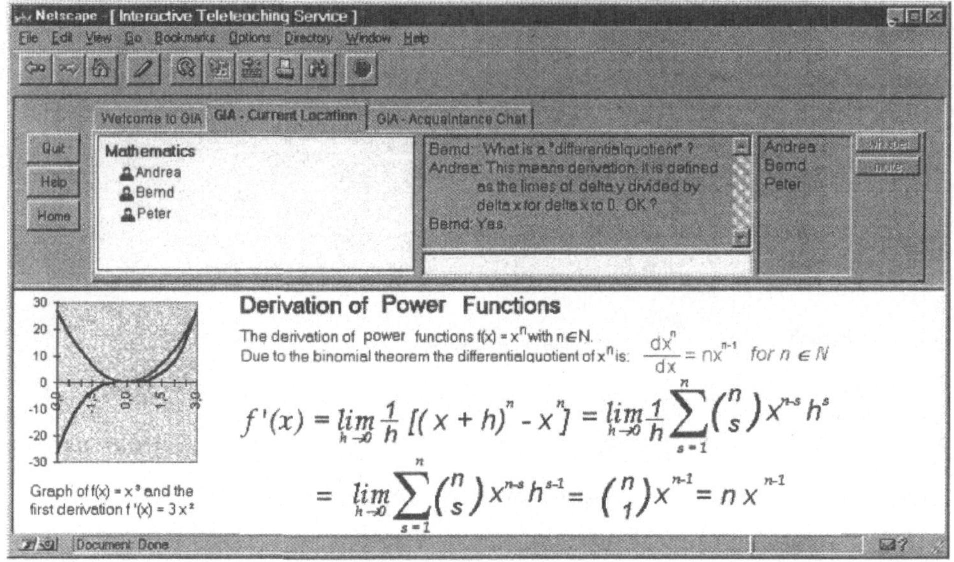

$$\frac{dx^n}{dx} = nx^{n-1} \quad for\ n \in N$$

$$f'(x) = \lim_{h \to 0} \frac{1}{h}\left[(x+h)^n - x^n\right] = \lim_{h \to 0} \frac{1}{h}\sum_{s=1}^{n}\binom{n}{s}x^{n-s}h^s$$

$$= \lim_{h \to 0}\sum_{s=1}^{n}\binom{n}{s}x^{n-s}h^{s-1} = \binom{n}{1}x^{n-1} = n\,x^{n-1}$$

Abb. 2: Die virtuelle Analysis-Gruppe

3. Dynamisches Gruppieren

Eine Idee für den Gruppierungsprozeß ist es, die Leute zusammenzubringen, die gleiche Interessen teilen, über ähnliches Wissen verfügen, oder die gleichen Lernprobleme und Fragen haben. Als Basis für die Verwirklichung dieser Idee, dient die Einführung von speziellen Mechanismen, im weiteren *Metriken* genannt. Diese Metriken erlauben eine flexiblere Realisierung der Wahrnehmung anderer Personen und garantieren damit eine unterschiedliche, dynamische, aber strukturierte Gruppierung. Sie geben dem Benutzer die Möglichkeit, von anderen zu profitieren und dies in Abhängigkeit von der eigenen Intention. Im Kontext des kooperativen Lernens bieten sie eine dynamische Auswertung, die auf der Metrik-Auswahl des Lernenden basiert. Dabei können für diese Auswahl auch verschiedene Metriken kombiniert werden. Wichtig ist, daß Metriken unabhängig von einer Applikation definiert werden und somit als Basis für unterschiedliche Anwendungen dienen können. Momentan sind drei Metriken für den Telelerndienst vorgesehen:

3.1 Die Raum-basierte Metrik

Die Dokumente des Telelerndienstes sind als vermaschtes Netz von HTML-Dokumenten aufgebaut und auf einer Website im Intranet abgelegt. Dies entspricht einem Graphen, der aus Pfeilen (eingebettete Links) und Knoten (Dokumente) besteht. Ein strukturierter Gruppierungsprozeß, der sich auf einer *raum-basierten Metrik* begründet, bringt die Benutzer zusammen, die jeweils ein Dokument innerhalb eines zu definierenden Teilgraphen anfordern. Dieser Graph wird bestimmt durch die Variable *Link-Distanz* (LD), die den kürzesten Weg zwischen einem ausgezeichneten Knoten (Knoten_spez) und einem weiteren Knoten

(Knoten_i) definiert. Dies bedeutet, daß die raum-basierte Metrik Überdeckungen im Graph sucht. Für diese Berechnung kann auf bekannte Graphenalgorithmen [7] zurückgegriffen werden. Es werden also zwei Benutzer (client_1, client_2) gruppiert, wenn das folgende Prädikat zu „wahr" ausgewertet werden kann:

```
Abstand (Knoten_spez(client_1), Knoten_i(client_2))<= LD
```

Die raum-basierte Metrik gruppiert also Benutzer, die sich durch ihre angefragten Dokumente in einer räumlichen Nähe zueinander befinden, wobei der Raum durch die auf meist inhaltlichen Kriterien basierende Verknüpfungsstruktur dieser Dokumente gegeben ist. Sie kann deshalb als mensch-zentriert und dokumenten-zentriert charakterisiert werden.

3.2 Die Semantik-basierte Metrik

Die *semantik-basierte Metrik* steht in engem Zusammenhang mit dem Inhalt der HTML Dokumente. Diese Metrik ist also sehr hilfreich zur Kontaktaufnahme mit anderen Benutzern oder Gruppen, wenn es um Interessen an Dokumenten mit ähnlichem Inhalt geht. Denkbar für die Realisierung dieses Ansatzes sind (1) die Einführung gewichteter Hyperlinks zwischen den Dokumenten, oder (2) die Extrahierung von inhalts-bezogener Information aus den Dokumenten. Die semantische Metrik kann als dokumenten-zentriert bezeichnet werden.

Zu (1): Der Pfeil zwischen zwei Dokumenten wird mit einem Gewicht versehen (gewichteter Hyperlink). Dieses Gewicht drückt die Intensität der inhaltlichen Beziehung zwischen den beiden Dokumenten aus. Ein hohes Gewicht impliziert dabei eine hohe Korrelation. Die Funktionsweise ist dabei die folgende: Gegeben sei ein Schwellwert W. Zwei Benutzer (client_1, client_2) werden nun miteinander gruppiert, wenn jeder Hyperlink, auf einem Weg zwischen den beiden angeforderten Dokumenten (Knoten_i, Knoten_j), einen Wert größer oder gleich dieses Schwellwertes hat. Die Werte können dabei vom Autor explizit zu jedem Link ins Dokument gesetzt werden oder halbautomatisch durch ein Analyse-Tool, das beispielsweise auf einem Keyword-Matching Algorithmus basiert, erzeugt werden. Es ist folgendes Prädikat auszuwerten:

```
Gewicht (Knoten_i(client_1), Knoten_j(client_2))>= W
```

Zu (2): Bei der *inhalts-basierten* Form der semantischen Metrik sollen Leute gruppiert werden, deren angeforderte Dokumente gleiche Themen adressieren. Um den relevanten Inhalt eines Dokumentes zu erfassen, sind Information Retrieval Algorithmen notwendig. Um diesem Ansatz die Charakteristik der NP-Vollständigkeit zu nehmen, könnte zum Beispiel im Hintergrund periodisch ein Suchalgorithmus laufen, der die Korrelationsinformation quasi statisch hält. Das folgende Prädikat muß zu „wahr" ausgewertet werden können, wenn zwei Benutzer gruppiert werden sollen:

```
Inhalt-match (Knoten_i (client_1), Knoten_j (client_2)) = 1
```

3.3 Die Metrik der Benutzer-Interessen

Bei der Metrik der Benutzer-Interessen steht eindeutig der Mensch im Vordergrund (mensch-zentriert). Hier findet der Gruppierungsprozeß anhand der Korrelation von Attributen, die dem Benutzer zugeordnet sind, statt. Attribute können dabei unterschiedlicher Natur und Ausprägung sein; zum Beispiel das Sprechen der gleichen Sprache, die Zugehörigkeit zur gleichen kulturellen Gruppe, eine gleiche berufliche Ausbildung, das Interesse an einem bestimmten Gebiet der linearen Algebra, und so weiter. Diese Attribute sind sehr unterschiedlicher Natur und lassen sich deshalb sehr schlecht zur Auswertung in eine Schema einordnen. Attribute können auf Client-Seite gesammelt und mit dem Server beim Browsen ausgetauscht werden. Das Prädikat, das dabei ausgewertet werden muß, ist dabei das folgende:

```
Int-Match (Attribut-Liste(client-1), (Attribut-Liste(client-2)) = 1
```

Für eine praktische Umsetzung sind zwei Wege denkbar: (1) Erstellung der Benutzer-Attributsliste durch Benutzerverfolgung, oder (2) Erstellung einer Liste durch Benutzerauswahl aus einer vordefinierten, eventuell erweiterbaren Tabelle.

Bei der Wahl einer Metrik oder einer Kombination spielen Komplexität und Implementierungsaspekte eine wichtige Rolle. Jeder Metrik gehen unterschiedliche Manipulationen auf Server-Seite voraus. Kombinationen von Metriken sind unter Umständen sinnvoll, da sie den Suchraum stark einschränken und dadurch die Berechnungskomplexität reduzieren. Weiterhin führen Metrik-Kombinationen zu einer gewünschten Limitierung der Gruppengröße; nur damit kann synchrones, kooperatives Lernen sinnvoll durchgeführt werden.

4. Möglichkeiten zur Benutzerinteraktion

Kapitel 3 beschrieb verschiedene Mechanismen, als Basis für eine dynamische, strukturierte Gruppierung. Diese können bei kooperativem Lernen, entsprechend der Präferenzen des Lernenden, eingesetzt werden. Kapitel 4 führt nun in verfügbare Features des Prototyps ein, die zur Unterstützung von Interaktion der Benutzer dienen. Zur Verfügung stehen Wahrnehmung eines Benutzers, Orts-Chat, Bekannten-Chat und kooperative Navigation.

Aufgabe der *Wahrnehmung eines Benutzers* ist es, die dazugehörigen URLs zu sammeln, und die entsprechende, benötigte Information zu generieren, um den Benutzer in einer 2D-Umgebung darzustellen (Abb. 1). Jeder Ort, der dabei einer logisch verknüpften Menge von URLs entspricht, wird in einem Raumplan (Abb. 1 c) repräsentiert. Zusätzlich wird jeder Lernende (Benutzer des Systems) einem Ort zugeordnet und visualisiert.

Das Feature *Orts-Chat* ermöglicht einem Benutzer eine Kommunikation mit anderen Benutzern, die sich in seiner virtuellen Umgebung befinden (Abb. 1 d). Wenn ein Benutzer den Raum wechselt, kann er die Gruppe wechseln, da die Gruppe dynamisch aus den momentanen Orten der Benutzer gebildet werden.

Im Gegensatz zum Orts-Chat, bei dem sich die Mitglieder in der Regel nicht kennen, kann das *Bekannten-Chat* für eine Kommunikation innerhalb einer geschlossenen sozialen Gruppe benutzt werden. Die Teilnehmerschaft ist dabei unabhängig vom momentanen Ort.

Für die Gruppe wird *kooperative Navigation* angeboten. Es erlaubt Mitgliedern zu einer aktuellen Seite eines anderen Mitglieds zu springen. Es ermöglicht geführte Touren als Navigations-Unterstützung für Benutzer, die den Service nicht kennen,. Alle Gruppenmitglieder werden automatisch synchronisiert, wenn der Gruppenlotse eine neue Seite anwählt.

Die obigen Features sind „hart verdrahtet" und können durch eine Anwendung der Metriken erheblich erweitert werden.

5. Architektur des Prototyps

Zur Überprüfung der Konzepte, wurde als Prototyp eine Client-Server Architektur, genannt GIA (Abb. 3), implementiert [4].

Der GIAServer verwaltet dabei alle anfallende Client Information, z.B. Name des Lernenden, sein momentaner Ort, Informationen über seine Gruppenzugehörigkeit, deren Mitglieder, etc. Weiterhin behandelt er die Funktionen Benutzer-Wahrnehmung und Bekannten-Chat. Der Server läuft als zusätzlicher Service auf einem Internet Server.

Konzeptuell besteht der GIAClient aus zwei Blöcken: Einer managt den Raumplan, realisiert die Wahrnehmung der Benutzer und visualisiert die Orte, der momentan den Service benutzenden Lernenden

Der zweite Client-Block realisiert die Chat-Funktionalität einer jeden Gruppe.

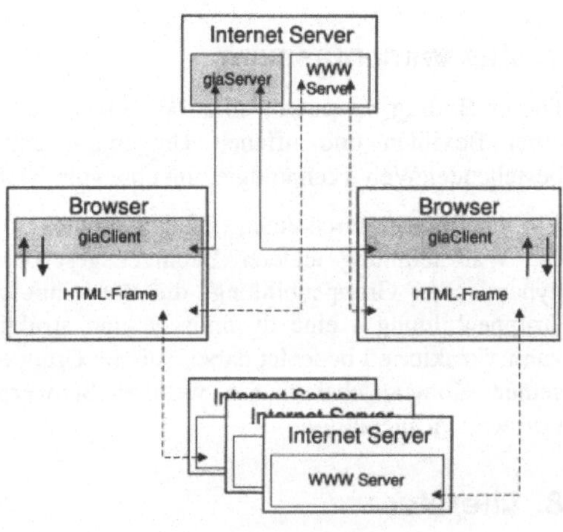

Abb. 3: Die Client-Server Architektur des GIA-Systems

6. Verwandte Arbeiten

Die Wahrnehmung der Benutzer ist eines der wichtigsten Kriterien für die Interaktion von Gruppen im Web [2]. Palfreyman und Rodden [6] haben ein Protokoll zur Realisierung von Benutzerwahrnehmung im Web veröffentlicht, aber sie adressieren dabei nicht den Gruppenaspekt.

Da neues Wissen und Entwicklungen ständig zunehmen, besteht für Organisationen, Gruppen ·und Einzelpersonen ein großer Bedarf das eigene Wissen und die eigene Expertise zu erweitern. Es gibt verschiedene synchrone und asynchrone Lernsysteme, die dies anbieten:

Das asynchrone System „Answer Garden 2" [1] stellt die zweite Generation einer Systemarchitektur zur Verfügung, die sich durch „organizational memory" und kollaborative Unterstützung charakterisieren läßt.

Bei „GestureCam" von Kuzuoka u.a. [3] handelt es sich um ein synchrones Lernsystem, das räumliche Zusammenarbeit über ein video-vermitteltes Kommunikationssytem erlaubt. „CLARE" [8] repräsentiert ein kollaboratives Lernsystem, das auch den Aufbau von Wissen erlaubt.

Verfügbare Lernsystemen zielen oft auf die Maximierung der Lerninhalte ab. Der Nachteil besteht darin, daß weder Gruppenunterstützung (findet nur in Form vordefinierter Lehrer-Schüler Beziehung statt), noch dynamische und strukturierte Gruppenbildung von Lernenden unterstützt wird.

7. Zusammenfassung

Dieser Beitrag beschreibt einen Web-basierten Telelerndienst, der gemeinsames Lernen in einer flexiblen und offenen Umgebung unterstützt. Die Implementierung basiert auf bestehender Web-Technologie, um eine gute Verbreitung zu ermöglichen.

Um das System durch einen sozialen Kontext zu erweitern, wurde ein Hauptschwerpunkt auf die Wahrnehmung anderer Benutzer gesetzt. Die Wahrnehmung bildet eine Basis zur dynamischen Gruppenbildung, die über innovative Mechanismen, in eine neue Art der Gruppenbildung - eine dynamische und strukturierte Gruppenbildung - überführt werden kann. Strukturiert bedeutet dabei, daß die Gruppenbildung auf die Belange des Lernenden und seinen Kontext abzielt. Ein weiterer Schwerpunkt im Prototyp unterstützt ausgewählte Formen der Interaktion.

8. Literatur

[1]   M.S. Ackerman, D.W. McDonald: Merging Organizational Memory with Collaborative Help. In: Proceedings, ACM 1996 Conference on Computer Supported Cooperative Work. Boston, MA, USA, 1996

[2]   G. Fitzpatrick, S. Kaplan, T. Mansfield: Physical Spaces, Virtual Places and Social Worlds: A study of work in the virtual. In: Proceedings, ACM 1996 Conference on Computer Supported Cooperative Work. Boston, MA, USA, 1996

[3]   H. Kuzuoka, T. Kosuge, M. Tanaka: GestureCam: A Video Communication System for Sympathetic Remote Collaboration. In: Proceedings, ACM 1994 Conference on Computer Supported Cooperative Work. Chapel Hill, NC, USA, 1994

[4]   P. Manhart, K. Schmidt, H. Ziegler: Group Interaction in Web-based Multimedia Market Places. In: Proceedings of the 31st Annual Hawaii International Conference on System Sciences, 1998, Vol. VII

[5]   K. Nakabayashi, Y. Koike, M. Maruyama: An intelligent tutoring system on the world wide web: Towards an integrated learning environment on a distributed hypermedia. ED-Media, 1995.

[6]   K. Palfreyman, T. Rodden: A Protocol for User Awareness on the World Wide Web. In: Proceedings, ACM 1996 Conference on Computer Supported Cooperative Work. Boston, MA, USA, 1996

[7]   U. Schöning: Algorithmen – kurz gefaßt. Spektrum Akademischer Verlag, 1997

[8]   D.Wan, P.M. Johnson: Computer Supported Collaborative Learning Using CLARE: the Approach and Experimental Findings. In: Proceedings, ACM 1994 Conference on Computer Supported Cooperative Work. Chapel Hill, NC, USA, 1994

Adressen der Autoren

Dr.-Ing. Karin Schmidt
Daimler Benz AG
Forschung Informationstechnik
Postfach 2360
89013 Ulm
Email:karin.schmidt@dbag.ulm.DaimlerBenz.COM

Dipl.-Inform. Johannes Bumiller
Daimler Benz AG
Forschung Informationstechnik
Postfach 2360
89013 Ulm
Email:bumiller@dbag.ulm.DaimlerBenz.COM

Dipl.-Inform. Peter Manhart
Daimler Benz AG
Forschung Informationstechnik
Postfach 2360
89013 Ulm
Email: manhart@dbag.ulm.DaimlerBenz.COM

Einsatz von Java Applikationen für das Organisationsdesign virtueller Unternehmen

Dipl.-Wirt. Inform. Marcus Ott und Carsten Huth
Lehr- und Forschungseinheit Wirtschaftsinformatik 2
Universität-GH Paderborn

Zusammenfassung

Organisationsstrukturen, die ausschließlich durch Hierarchien, Bürokratie und viele Entscheidungsebenen gekennzeichnet sind, erscheinen nicht länger geeignet, den vieldiskutierten Umweltanforderungen nach Reaktionsschnelligkeit gerecht zu werden. Netzorientierte Informationstechnologie und innovative Kommunikationsformen eröffnen Wege, die zusammen mit einer Neuordnung von Strukturen und Aufgabenausführung dazu beitragen können, bewegliche Organisationsformen zu realisieren.

Dieser Bericht beschreibt die innovative Kombination zweier in sich grundlegend neuer Technologien: Java Programming und Groupware. Dabei werden zwei Bereiche adressiert. Zum einen werden die akuten Anforderungen erörtert, die heutige Organisationen zu Strukturformen zwingen, die virtueller Art sind, die immer stärker den Workgroupaspekt in den Vordergrund stellen und die eine vernetzte Arbeitsweise unabdingbar machen. Zum anderen wird sich auf Basistechnologien konzentriert und aufgezeigt wie im Rahmen des GroupOrga Projektes z.B. mit Java-basierten Werkzeugen Lösungen für die sich stellenden Anforderungen erarbeitet werden.

Teil I. Organisationsdesign in virtuellen Gemeinschaften
Teil II. Java Applikationen für plattformübergreifende Interaktion

I   Organisationsdesign in virtuellen Gemeinschaften

I.1   Szenario

Konventionelle Organisationskonzepte sind seit hunderten von Jahren in unseren Unternehmen verankert und haben sich über lange Zeit bewährt. Da sich jedoch die Konkurrenzsituationen verschärfen und der Faktor Beweglichkeit einen immer größeren Stellenwert einnimmt, müssen sich Unternehmen darauf einstellen, schneller auf ihre Umweltveränderungen zu reagieren und Anforderungen des Marktes flexibler gerecht zu werden. Stärkerer Wettbewerb, Globalisierung der Märkte und gesteigerte Erwartungen der Kunden wirken sich als zunehmender Druck auf Unternehmen und ihre starren Strukturen aus. Ein ganz besonderer Effekt zeichnet sich durch eine anwachsende Geschwindigkeit im Bereich der technologischen Veränderungen und Neuerungen ab, was zu einer Neuordnung der Wettbewerbsfaktoren führt: *Zeit* spielt den mitentscheidenden Faktor.

Organisationsstrukturen, die sich durch Hierarchien, Bürokratie und viele Entscheidungsebenen auszeichnen, erscheinen ungeeignet, diesen Anforderungen nach Reaktionsschnelligkeit gerecht zu werden. Informationstechnologie (IT) eröffnet Wege, die zusammen mit einer Neuordnung von Strukturen und Aufgabenausführung dazu beitragen kann, bewegliche Organisationsformen zu realisieren. Im Bereich des *Computer Supported Cooperative Work* (CSCW) werden vielfache Ansätze diskutiert, dieses Ziel zu erreichen. Im

Verständnis von *Business Process Reengineering* (BPR) und *Workflow Management* werden dabei Lösungen untersucht, die explizit die Geschäftsprozesse und deren Verarbeitungszeit ansprechen und eine Verbesserung in der Reaktion auf Umweltveränderungen bezwecken. Jedoch wird es schwierig und zuweilen unmöglich sein, traditionelle Unternehmen dazu zu bewegen, innovative Geschäftsprozeßstrukturen anzunehmen, ohne gleichzeitig eine Veränderung der *zugrundeliegenden Organisationsstrukturen* zu erwirken.

Dieser Beitrag erörtert eine Zahl von technologiebasierten Designansätzen und -technologien, von denen einige ähnlich, andere verschieden von traditionellen aufbaustrukturellen Designvariablen sind. Ein solches IT-unterstütztes Design kombiniert u.a. virtuelle organisatorische Komponenten, Gruppenstrukturen und elektronische Prozeßabwicklung. Es wird ein Ansatz beschrieben, der darauf ausgerichtet ist, die Strukturen virtueller Gemeinschaften zu entwickelt, die dabei aber mehr aus Wissensknoten und inter-organisationalen Netzwerken besteht als aus realen Lokationen und Mitgliedern.

Strukturierung virtueller Gemeinschaften erscheint dabei zunächst als Widerspruch in sich. Was jedoch aufgezeigt wird ist die notwendige Vorstrukturierung und Bestimmung von Grenzen innerhalb virtuellen Infrastrukturen, deren Feinentwicklung den Entscheidungen der Wissensträger in den einzelnen virtuellen Gemeinschaften obliegt.

I.2 Workgroups in virtuellen Gemeinschaften

Katzenbach und Smith ([9], S. 70) definieren eine Workgroup als eine kleine Gruppe von Personen, deren Fähigkeiten einander ergänzen und die alle daran beteiligt sind, ein gewisses gegebenes Ziel zu erreichen. Die Personen in einer solchen Workgroup haben ein gemeinsames Bestreben, einen kooperativen Arbeitsansatz und sie ziehen sich gegenseitig zur Rechenschaft. In einer früheren Untersuchung ([17]) wurde der Trend zur Arbeit in Workgroups als angetrieben von schnellebigen Märkten und einem Trend zu flacheren Hierarchien diagnostiziert. Hinzu kommen als Gründe teambasierte Bewertungsansätze oder beispielhafte Berichte in denen Unternehmen durch eine Konzentration auf ihre Workgroups Wettbewerbsvorteile erringen konnten.

Workgroups stellen eine weitere Form der Arbeitstrukturierung dar, die parallel zu - und nicht als Ersatz von - traditionellen Konzepten von Hierarchie und Leistung verstanden werden sollte. Workgroups können sich dabei auch in die formalen Strukturen und Prozesse integrieren und sie fördern. Hierarchische Strukturen und die darauf begründeten Prozesse sind wesentlich in großen Organisationen - durch die anwachsende Verbreitung von Workgroups sind Hierarchien jedoch nicht unvermeidlich in Frage gestellt. Der Einsatz von Workgroups ist im Gegenteil eine Möglichkeit, die bestehenden strukturellen Grenzen zu überbrücken und die Kernprozesse einer Organisation zu unterstützen. Werden Workgroups als *Ersatz* traditioneller Strukturen gesehen ist ihr wirkliches Potential mißverstanden.

Besonderes Interesse an Workgroups und gruppenbasierten Organisationsformen erwächst aus der Koexistenz von Organisationsrestrukturierung und dem gleichzeitigen Einsatz von produktiver, gruppenorientierter Informationstechnologie, wie z.B. dem World Wide Web (WWW) oder Groupwareplattformen. Eine wachsende Bereitschaft für eine Umstrukturierung konventioneller Organisation kann festgestellt werden, die einhergeht mit der Einführung innovativer Informationstechnologie (vgl. [20], S. 10). Diese Kombination erlaubt eine neue Form und Qualität der Interaktion von Workgroups: Computergestützte gemeinsame Arbeitsräume und Teamwork in virtuellen Arbeitsgemeinschaften.

I.3 Virtuelle Gemeinschaften und Organisationsdesign - ein Gegensatz?

Mit der Unterstützung durch Informationstechnologie wird die bisher grundlegende Voraussetzung von Organisationstheorie und -praxis mehr und mehr ungültig. Diese traditionsgemäß geht davon aus, daß man Personen und Abteilungen örtlich gruppieren muß, um Koordination und Überwachung zu verwirklichen und daß eine Entscheidung zwischen Zentralisation und Dezentralisation notwendig ist. Im Falle des bisher notwendigen persönlichen Verbindungsaufbaus ist es durch IT so, daß z.B. elektronische Post (e-mail im Internet/WWW) oder Groupware nun diese Rolle übernehmen können und damit lokal gebundene Projektgruppen und Verbindungspersonen obsolet macht. Im Gegensatz zur physischen Präsenz erlaubt IT Einsatz den Aufbau virtueller Organisationsstrukturen und Gemeinschaften. Virtuelle[1] Gemeinschaften (VG) hatten ihren Ursprung zu einer Zeit, als man begann das Potential von IT für die Arbeit an verteilten Arbeitsplätzen zu entdecken. Für fast jedes Unternehmen, das kein materielles Gebrauchsgut herstellt, ist eine mögliche Unternehmensform die der Kombination von zeitlich und örtlich unabhängigen Akteuren oder Workgroups. In einem solchen Szenario ist eine Organisationsform mit herkömmlichen Strukturen und den dazugehörigen persönlichen Treffen nicht länger notwendig.

Der Gebrauch des Begriffes 'virtuell' variiert stark. In den heutigen IT Konzepten von Netzwerkorganisationen und in der Literatur werden unterschiedliche Interpretationen und Schwerpunkte von VG gegeben: Davidow und Malones "Virtual Corporation" zum Beispiel, bezieht sich ausschließlich auf die äußere Form einer Organisation, wenn sie VG als Objekte ohne spezifische Konturen und sich beständig ändernder Schnittstellen zwischen Organisation, Partnern und Kunden beschreiben [6, S. 15]. Andere Autoren konzentrieren sich auf unterschiedliche Typen der Kooperation zwischen gleichberechtigten Partnern. Ein engeres Konzept wird von Byrne angeboten, der festlegt "... the virtual corporation is a temporary network of independent companies [...] linked by information technology to share skills, costs, and access to one another's markets" [3, S. 37]. Wieder andere Autoren verstehen die Idee virtueller Gemeinschaften nicht nur als Erklärung interorganisationaler Beziehungen, sondern besonders auch intraorganisationaler Abhängigkeiten. Arnold et al. [2] haben den Begriff *virtuelle Organisation* im Detail untersucht und die Kennzeichen dieser Form durch Abgrenzung von Joint Ventures, Kartellen, Großkonzernen, Strategische Allianzen und ähnlichen Organisationsformen abgesteckt. Der Fokus dieser Arbeit ist nicht, eine weitere Definition virtueller Gemeinschaften zu entwickeln, da dies ausreichend erfolgt ist[2]. Für den

---

[1] Der Begriff "virtuell" wurde zuerst im Zusammenhang mit IT für die Beschreibung eines quasi unbeschränkt großen logischen Speichers für Computer verwendet. Dabei wurde ein Programm in sog. 'Seiten' aufgeteilt und nur die notwendigen Seiten wurden in den physischen Speicher geladen. Eine virtuelle Gemeinschaft nutzt IT, um wie eine klassische Gemeinschaft zu operieren.

[2] Einige Literaturquellen für weitere Studien die einen Überblick über virtuelle Arbeitsformen geben sind [2], [3], [4], [5], [10], [11], [12], [14], [18], [21].

Zweck dieser Fachtagung basiert der hier dargestellte Ansatz auf Definitionen aus [21, S. 383] und [2, S. 10].

- Damit wird eine virtuelle Gemeinschaft verstanden als eine freiwillige Kooperation verschiedener rechtlich unabhängiger Aufgabenträger unterschiedlicher Aggregation (ganze Organisationen, einzelne Abteilungen, Workgroups, einzelnen Personen, etc.).
- Diese Aufgabenträger suchen, basierend auf einem gemeinsamen Verständnis ihrer Regeln, ein gewünschtes Ergebnis zu erzielen.
- Alle kooperierenden Partner stellen dabei ihre Ressourcen, Kernkompetenzen, sowie ihr Know-how und ihre Fähigkeiten zur Verfügung, um schneller zu reagieren, flexibler zu sein und ggf. international agieren zu können.
- Wenigstens ein Partner einer solchen virtuellen Gemeinschaft repräsentiert die Gemeinschaft nach außen und hat i.d.R. auch die strukturierende Verantwortlichkeit.
- Die Partner sind dabei über moderne Informations- und Kommunikationstechnologie verbunden.

Basierend auf diesem Verständnis geht der hier dargestellte Ansatz davon aus, daß Teile des klassischen Aufbauorganisationsdesigns weiterhin fundamentale Managementaufgaben in VG sind und bleiben. Mehr noch soll herausgestellt werden, daß *gerade* VG einer koordinierenden und strukturierenden Komponente bedürfen, um ein Entgleiten der Partner zu verhindern und die virtuelle Aufbaustruktur zu dokumentieren. Klein definiert dieses Design organisatorischer Architekturen ebenso als eine elementare Managementaufgabe: "Auch virtuelle, flexible Organisationen bedürfen eines Mindestmaßes an Struktur. Es sind daher grundlegende Organisationsprinzipien zu bestimmen und die Rechte und Verantwortungen der Organisationseinheiten und der Mitarbeiter zu klären." [10, S. 313].

Die Notwendigkeit IT einzusetzen, wie die hier dargestellte zur Koordination virtueller Gemeinschaften, wird insbesondere unter dem Gesichtspunkt deutlich, den Gurbaxani und Whang hervorheben: Virtuelle Gemeinschaften können eine enorme Größe erreichen, während die teilnehmenden Partner vergleichsweise klein sein können [8]. Dem externen Partner (genauso wie auch den internen Partnern) gegenüber präsentiert sich die virtuelle Gemeinschaft als eine transparente Entität von außergewöhnlicher Komplexität. Dieser Umstand bedingt umsomehr den Einsatz von IT zur Dokumentation und Strukturierung virtueller Gemeinschaften.

Weiterhin scheint es so, daß für eine gewisse Zeit die Metapher traditionell strukturierter Organisation weiterhin das Verständnis der Mitglieder VG beherrschen wird - auch wenn sie eine virtuelle Form beschreibt. Zum gegebenen Zeitpunkt ist es eher unrealistisch, reine VG vorzufinden, sondern eher eine Kombination realer Organisationsstrukturen und virtueller Komponenten.

Im folgenden Kapitel wird ein verteilter, evolutionärer Designprozeß für Organisationsstrukturen beschrieben. Dieser Prozeß gestaltet sich als ein Vorgehen auf mehreren Ebenen, der alle Mitglieder einer VG in den dauerhaften Designprozeß einbezieht.

Kapitel I.I.5 setzt darauf auf und führt in die CSCW Umgebung *GroupOrga* ein, welche einen kritischen Faktor im Designprozeß VG abdeckt: Jederzeit und für jeden aktuelle und verfügbare Information bereitzustellen, über Strukturen, Prozesse, Partner und

Verantwortlichkeiten sowie über Know-how, Fähigkeiten und Dienste, die von den Partnern des virtuellen Netzwerkes angeboten werden.

I.4 Interaktive Workgroups im World Wide Web

Um Organisationen an veränderte Umweltbedingungen anzupassen, sind die oben angedeuteten Prozesse struktureller Veränderung notwendig. Wie in [16] gezeigt, kann Groupware eine entscheidende Technologiebasis für diese Prozesse sein, da sie neue Modi für Arbeitsteilung innerhalb und zwischen den beteiligten Partnern anbietet. Weiterhin können synchrone und asynchrone Kommunikationssysteme die Selbstkoordination der mitwirkenden Partner unterstützen und Koordination in Form von Formalismen und hierarchischen Entscheidungen ersetzen. Jedoch bietet nicht nur Groupware diese Möglichkeiten und ein immer stärker zu verzeichnendes Zusammenwachsen von Groupware Technologie und dem WWW veranlaßt dazu, die Interaktion von Workgroups auch oder gerade im WWW zu betrachten.

Für Gene Phifer, Forschungsdirektor bei Gartner Group, Inc. in Stamford, Conn., wird die Groupwarearena in Zukunft am meisten durch WWW basiertes Networkcomputing beeinflußt werden: "Network computing will change groupware because of different access mechanisms, different paradigms for access, the whole bringing in the network computer model, which is really the next step in the client/server model" (zitiert in [7]).

Der folgende Abschnitt beschreibt einen allgemeinen Designprozeß organisatorischer Strukturen, wie er durch virtuelle Gemeinschaften mit Technologieunterstützung der beiden genannten Formen möglich wäre.

I.4.1 Ein Designprozeß durch Workgroups

Vorstellung des im GroupOrga[3] Projekt dargestellten Ansatzes ist es, daß der von Workgroups initiierte und geführte Strukturdesignprozeß virtuelle Gemeinschaften hervorbringt, die selbstorganisierende Systeme sind. Ein solcher Prozeß ist ein bewußter, dauerhafter Problemlösungs- und Planungsprozeß, der ein Design hervorbringt, welches sich ständig weiterentwickelt und verändert.

---

[3] GroupOrga (*Group*ware basiertes *Orga*nisationsdesign) ist ein von Dipl.-Wirt. Inform. Marcus Ott geleitetes Forschungsprojekt der Lehr- und Forschungseinheit Wirtschaftsinformatik 2 von Prof. Dr. Ludwig Nastansky an der Universität Paderborn. http://fb5www.uni-paderborn.de/winfo2/GroupOrga

Abb. 1 zeigt einen solchen andauernden Organisations- und Strukturierungsprozeß. In diesem Prozeß entwickeln sich innerhalb einer aktuellen Struktur neue virtuelle Gemeinschaften (①). Mit der Zeit werden einige dieser Gemeinschaften formalisiert und entwickeln sich zu festen und stabilen Punkten in einem ansonsten wenig strukturierten Umfeld (②). Während sich die neu herausbildenden Strukturen verfestigen und verbinden (③),

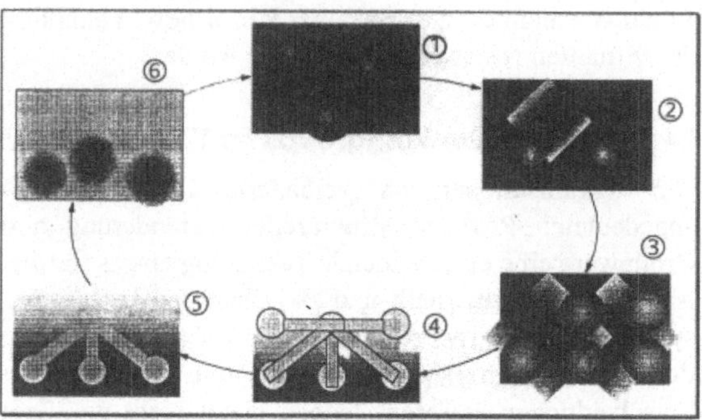

Abb. 1: Gruppendesignprozeß für Organisationsmodellierung

lösen sich alte und bisher bestehende selber wieder auf (④) - in diesem Stadium ist es schwierig, alt von neu zu unterscheiden (⑤). Nach und nach bilden sich die alten Strukturen als Inseln im Neuen heraus (⑥) und lösen sich schließlich auf, wenn sie nicht mehr funktionell sind (Stadium ① in einem neuen Kreislauf).

Dieser Gruppenprozeß zum Design virtueller Gemeinschaften schließt jeden mit ein, der Interesse an Organisation zeigt, also nicht nur die direkten Mitglieder. Er basiert auf einer andauernden (computerunterstützten) Absprache der Beteiligten über ihre Tätigkeiten, über die Rollen oder Funktionen, die sie ausfüllen und darüber, wie ihre Rollen mit anderen Rollen interagieren. Diese Absprachen werden zum Beispiel in angespannten Situationen zunehmen, werden aber auch sonst immer vorhanden sein, und wenn es nur darum geht, eine Besonderheit zu einem Routinevorgehen zu besprechen.

Die Technologie, um diesen Vorgang zu unterstützen, ist eine computerunterstützte Kommunikationsumgebung, die es Personen ermöglicht, ihre Rollen und Interaktionen nach außen zu verdeutlichen. Die unterliegende Prämisse der Technologie und des Gruppenprozesses ist es, die verschiedenen Ansichten über die Rollen und Interaktionen öffentlich zu machen. Ein Hauptvorteil davon ist, daß andere Beteiligte sehen können, was passiert und ihre Informationen und Anschauungen mit einbringen können. Durch dieses Vorgehen erkennen die Beteiligten die Zusammenhänge zwischen ihren und anderen Tätigkeiten und können somit zum Gruppenprozeß beitragen.

I.4.2 Aufbau gemeinsamen Wissens durch virtuelle Arbeitsplattformen

Der dargestellte Designprozeß basiert auf einenm andauernden und evolutionären Vorgehen, das kontinuierlich zur Ansammlung von Informationen über die Zusammenhänge, Verantwortlichkeiten zwischen und Fähigkeiten von Mitgliedern der virtuellen Gemeinschaften beiträgt. Dieser Prozeß sammelt organisatorisches Wissen über Strukturen an, transformiert es ggf. und stellt es wieder zur Verfügung. Demzufolge bietet ein solcher Prozeß eine Ausgangsbasis für den Aufbau gemeinsamen Wissens in virtuellen Gemeinschaften und beschreibt, wie die Mitglieder am Designprozeß teilhaben.

Unterschiedliche Ansätze betrachten die Aquisition gemeinsamen Wissens und die Form der Darstellung dieses gemeinsamen Wissens. Yates [22] betrachtet Dokumente als Wissensträger während Ackerman [1] sich auf Informationstechnologie konzentriert. Stein [19] betrachtet ebenso die Integration von Informationssystemen in virtuellen Gemeinschaften zur Ansammlung organisatorischen Wissens.

Strukturelles Wissen wird in VG durch Anpassung aufgebaut; es mag verglichen werden mit organisationalem Lernen. Dabei wird Wissen über positive und negative Formen organisatorischer Struktur durch Erprobung verschiedener Formen in gegebenen Umgebungen über Zeit angesammelt. Erst ein kontinuierlicher Veränderungsprozeß wie in Abb. 1 dargestellt erlaubt es virtuellen Gemeinschaften sich immer wieder optimal anzupassen. Die Implementierung und Unterstützung durch CSCW Umgebungen konstituiert dabei eine Form organisatorischen Wissens. Einmal eingesetzt erlaubt CSCW Technologie die Herausbildung bestimmter Strukturen zwischen den Mitgliedern virtueller Gemeinschaften.

I.4.3   Top-level und Bottom-level Design

Die CSCW Informationsumgebung, die hier konzeptioniert wird, basiert auf einer Client-Server Architektur und einem verteilten Directorymodell. Die Verteilung des organisatorischen Wissens kann dabei auf eine beliebige Zahl von Informationssystemen innerhalb des virtuellen Netzwerkes geschehen, was eine außergewöhnliche Skalierbarkeit der Wissensumgebung virtueller Gemeinschaften erlaubt.

Aufgrund einer Vorgabe, welche Teilinformation auf welchem Knoten im Netzwerk vorzufinden ist, wird ein verteiltes Design und Administration des Strukturwissens denkbar. In einem solchen Szenario ist eine zentrale Authorität (möglicherweise ein *Broker*) für koordinierende und strukturelle Information verantwortlich, während die dezentral angesiedelten Partner ihre detaillierten, strukturellen Informationen zur Verfügung stellen.

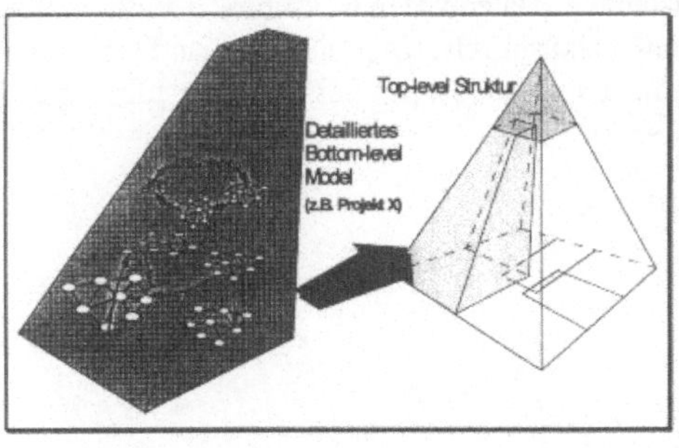

Abb. 2: Top-level und Bottom-level design

Der zentrale Punkt dieser Idee ist die Verwaltung solcher innovativen Netzwerke virtueller Gemeinschaften auf mindestens zwei Designebenen: Die erste Ebene, die Top-level Design genannt werden könnte, untersteht der Verantwortlichkeit der Koordinatoren virtueller Gemeinschaften - in traditionellen Strukturen würde diese Funktion als *Senior Management* bezeichnet. Die hier bezeichnete Gruppe ist mit der Rahmenbildung beschäftigt und legt die Infrastruktur aus Ressourcen, Hierarchien und Managementregeln fest. In Abb. 2 ist dies durch die Kuppe der Pyramide symbolisiert. Diese vorgegebenen strukturellen Elemente werden dann von den individuellen Partnern innerhalb der VG herangezogen, um die zweite Ebene des Strukturdesigns zu verwirklichen. Diese mit Bottom-level Design zu bezeichnende Tätigkeit ist durch Selbstorganisation gekennzeichnet. Dabei ist diese selbstorganisierende Schicht nicht auf eine Ebene unterhalb der zentralen Komponente (Top-level Design) beschränkt, sondern kann vielfache Ebenen umfassen. [15] untersucht die beiden möglichen in Abb. 2 illustrierten Designebenen vertiefend.

I.5 Eine CSCW Umgebung für die Interaktion virtueller Gemeinschaften

Der Designprozeß virtueller Gemeinschaften kann mit dem traditioneller Organisationsstrukturen verglichen werden [12, S. 63], d.h. dem Design der Geschäftsprozesse und der zugrundeliegenden Strukturen. Anstatt jedoch einzelne Personen vorzufinden, die die Aufgabe der Organisationsmodellierung durchführen, werden in einer virtuellen Gemeinschaft - unterstützt durch IT - alle Mitglieder am Designprozeß beteiligt sein. Nichtsdestotrotz müssen aber Prozesse und Strukturen offengelegt werden, um die Verantwortlichkeiten der einzelnen Mitglieder zu dokumentieren.

Technologisch basiert der GroupOrga Ansatz auf einer Zahl von interagierenden Komponenten. Dazu gehören ein Organisationsmodel, das GroupOrga Enterprise Information Management Model (GEIMM), ein elektronisches Organisationshandbuch in Form einer Organisationsdatenbank, das allen Mitgliedern der virtuellen Gemeinschaft zur Verfügung steht, und computerbasierte Werkzeuge. Diese Werkzeuge werden von allen Mitgliedern der VG eingesetzt, um die Metastruktur der Gemeinschaft zu modellieren und um später in der verteilten Umgebung die Ausformulierung auf der Ebene des Bottom-level Design zu verwirklichen. Sowohl das elektronische Organisationshandbuch, als auch die computerbasierten Werkzeuge mögen in standardisierten Umgebungen wie z.B. dem WWW eingesetzt werden, um die Strukturen, Kompetenzen und Fähigkeiten der Mitglieder von VG zu dokumentieren und darüber zu informieren.

Abb. 3 zeigt schematisch die integralen Komponenten der GroupOrga Architektur: Die verteilte Groupwareplattform (1) erlaubt es, das gemeinsame Wissen über Strukturen verteilt zu administrieren und jederzeit Zugriffsrechte und Sicherheitskonzepte zu definieren. Darauf aufbauend existieren gegenseitig konsistente Datenbestände, zum einen für (aufbau-)strukturelle Informationen (2), d.h. das elektronische Organisationshandbuch, und zum anderen ein Repository für Prozeßmanagementsysteme (3), mit dessen Unterstützung die zugehörigen Prozesse gesteuert und verwaltet werden.

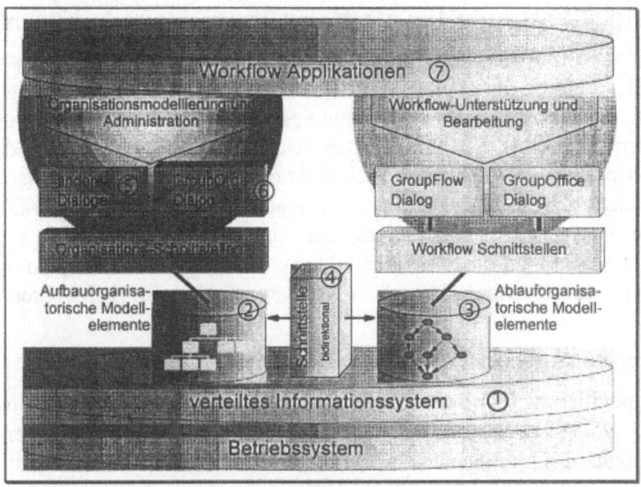

Abb. 3: Schematische Darstellung der GroupOrga Architektur

Eine Interface-Komponente zum Datenabgleich (4) erlaubt es, Informationen zur Organisationsstruktur verteilt über das Netzwerk direkt vom Organisationsdirectory zu operativen Systemen, z.B. für Workflowmanagement, zu transportieren. Basierend auf diesen beiden Pfeilern der Architektur finden sich die eigentlichen Anwendungen für Geschäftsprozesse, Bürotätigkeiten usw. innerhalb der virtuellen Gemeinschaften (7). Modifikationen und Aktualisierungen des gemeinsamen Wissens über Organisationsstrukturen wie sie im Aufbaurepository (2) festgehalten sind, können entweder direkt über das Datenbank-Frontend des Organisationshandbuches erfolgen (6), oder mit Unterstützung einfach zu handhabender, graphischer Adminstrationswerkzeuge (5).

Ein solches Werkzeug, der *GroupOrga Organization Modeler* wurde im Projekt als C++-Applikation entwickelt und befindet sich inzwischen im operativen Einsatz zur Unterstützung in verschiedenen Workflow Systemen. Die aktuelle Weiterentwicklung sowohl auf technischer und technologischer Seite in Form von *Java Programming*, als auch die konzeptionelle Erweiterung der Einsetzbarkeit und des Anwendungsbereiches dieses Werkzeugs sind Inhalt des zweiten Teils dieses Berichtes.

II   Java Applikationen für plattformübergreifende Interaktion

Aus den zuvor geschilderten konzeptionellen Anforderungen ergeben sich vielschichtige technische Anforderungen. In Kapitel I.I.5 wurden einige der Werkzeuge angesprochen, die aus konzeptionellen Überlegungen des GroupOrga Projekts entstanden sind und derzeit im Einsatz sind. Im Kapitel II.II.2 wird erörtert, welche Anforderungen mit den bisherigen softwaretechnologischen Möglichkeiten noch nicht erfüllt werden konnten. Nachfolgend wird verdeutlicht, welche Chancen sich durch Java ergeben (Kapitel II.II.3) und wie diese im Bezug auf die Organisationsmodellierung für virtuelle Organisationen genutzt werden können. Nach einer Sicht auf die Synergieeffekte verschiedener Technologien (Kapitel II.II.4) wird die prototypische Umsetzung der neuen Werkzeuge in Kapitel II.II.5 beschrieben.

II.1   Variierende Anwenderklassen in der Organisation

Organisationsmodellierung wird in Unternehmen von verschiedenen Anwenderklassen durchgeführt. Aus den verschiedenartigen Anforderungen der Anwender ergibt sich ein Nutzungskontinuum, das im folgenden beschrieben und in Tab. 1 illustriert wird.

Die erste Position einer solchen Skala nimmt der Endanwender ein, der das Bedürfnis hat, auf Knopfdruck Informationen über die aktuelle Organisationsstruktur zu bekommen und seine eigenen Daten bezüglich seines Know-how und seiner Fertigkeiten zu pflegen. Dieser Bereich betrifft jeden Mitarbeiter eines Unternehmens, insbesondere diejenigen, die nicht oder im Moment nicht in andere Organisationsdesignprozesse integriert sind. Der nächste Anwendertyp der Skala umfaßt Unternehmensmitglieder, die nur sporadisch Anpassungen an ihrem Bereich der Organisationsstruktur vornehmen - dies sind beispielsweise Abteilungsleiter oder andere Mitarbeiter im operativen Management. Ein weitere Ausprägung auf der Skala faßt die Mitarbeiter zusammen, die regelmäßig Änderungen an der Organisationsstruktur vornehmen, hier sind z. B. Mitarbeiter mittlerer Managementebenen zu nennen. Der letzte Bereich umfaßt z. B. die Mitarbeiter in Bereichen des strategischen Managements, die häufig Änderungen vornehmen. Hierunter sind auch die heute noch mit „Organisatoren" bezeichneten Stellen einer Organisation zu fassen.

Informierender Zugriff	*Sporadische Strukturanpassungen*	*Häufige Änderungen*	*Regelmäßige intensive Änderungen*
• *Information auf „Knopfdruck"* • *Pflege der eigenen organisatorischen Daten*	*Sporadische Organisations-strukturanpassungen und -änderungen*	*Einheitenübergreifend regelmäßige Änderungen und Planung*	• *Regelmäßige Gestaltung, Analyse und Bericht* • *Gestaltung von Grund auf*
Endanwender ◄——► Administrator			

Tab. 1: Variierende Anforderungen unterschiedlicher Anwenderklassen

II.2   Anforderungen an IT-Systeme zur Organisationsmodellierung

Im Vergleich der konzeptionellen Anforderungen, die weiter oben dargestellt wurden, mit der derzeitigen technischen Umsetzung ergibt sich in der Differenz eine Menge von Anforderungen, von denen wieder ein Anteil mit Hilfe von neuen technischen Möglichkeiten realisiert werden kann, während einige der genannten Anforderungen derzeit nur schwierig technisch umsetzbar bleiben. In diesem Kapitel wird die Menge der Anforderungen beschrieben, deren Lösungsansätze mit derzeit neuen Softwaretechnologien umgesetzt werden können.

Organisationsmodellierung durch alle Organisationsmitglieder

Die Beteiligung jedes Mitglieds eines Unternehmens an der Gestaltung der Organisationsstruktur impliziert, daß dies von jedem Rechnerarbeitsplatz eines Unternehmens möglich sein sollte (vgl. [16], Kap 5.1). In den meisten bestehenden Organisationen findet man heute aus verschiedensten Gründen heterogene Hardware-Plattformen und Betriebssysteme. Als Anforderung für ein Werkzeug zur Organisationsmodellierung läßt sich daraus ableiten, daß dieses unabhängig auf einer möglichst großen Anzahl von Hardware-Plattformen funktionsfähig sein sollte. Dieser Aspekt wird in Bezug auf virtuelle Organisationen sogar noch verstärkt. In dem Szenario einer virtuellen Organisation schließen sich mehrere klassische Organisationen für einen begrenzten Zeitraum zu einer Kooperation zusammen. Daher müssen nicht nur intraorganisationale sondern auch interorganisationale Organisationsmodellierungsprozesse unterstützt werden. Während einige Organisationen intern Standards für IT-Systeme einhalten, kann dieses für die verschiedenen Teilorganisationen einer VG nicht vorausgesetzt werden.

Kurze Rüstzeiten

Ein wesentliches Ziel einer virtuellen Organisation ist, schnell handlungsfähig zu sein. Sie kann durch eine Idee von mehreren unabhängigen Partnern oder durch eine marktwirtschaftlichen Entwicklung entstanden sein, die zum Zusammenschluß von mehreren Teilorganisationen führte. Ein Werkzeug zur Gestaltung von Aufbaustrukturen virtueller Organisationen sollte diesem Anspruch ebenfalls genügen, d. h. eine sehr kurze Rüstzeit haben. Informationstechnologisch betrachtet bedeutet dies, daß ein Installationsaufwand soweit wie möglich vermieden werden sollte und daß der Schulungsaufwand für die Anwender durch eine entsprechende Gestaltung der Software soweit wie möglich zu minimieren ist.

Reduzierung des Wartungsaufwands

Besonders Anwender der beiden ersten Abschnitte der Skala in Tab. 1 (*informierender Zugriff* und *sporadische Strukturanpassungen*) nutzen Organisationsmodellierungssoftware nur selten. Weiterhin besteht bei herkömmlichen Softwaretechnologien Wartung im allgemeinen darin, periodisch neue Versionen der Software zu installieren. Aus der Praxis ist bekannt, daß die Installation einer neuen Software auf allen Computerarbeitsplätzen einer Organisation einen hohen finanziellen Aufwand darstellt. Insbesondere für diese beiden genannten Teile der Skala von Anwendern steht deswegen der zusätzliche Nutzen durch den Einsatz der IT-gestützten Organisationsmodellierung dem hohen Aufwand der Pflege der

Softwareinstallationen entgegen. Eine erhebliche Verminderung dieses Wartungsaufwands könnte diesen betriebswirtschaftlichen Konflikt auflösen. Damit könnte ein Durchbruch für die GroupOrga Konzeption im Bereich des Bottom-level Design erreicht werden. Im nächsten Kapitel wird gezeigt, wie dieses Ziel unterstützt durch Java-basierte Organisationsmodellierungswerkzeuge erreicht werden kann.

Arbeitsplatzunabhängigkeit

Weiterhin gehört zu den Charakteristika eines virtuellen Unternehmens, daß Teams für kurzfristige Projekte flexibel zusammengestellt werden. Mitarbeiter eines Unternehmens können damit nicht von über lange Zeit stabilen Lokationen ihrer Arbeitsplätze ausgehen. Ein solcher Arbeitsplatz kann sich innerhalb eines Standorts ändern, aber auch ein projektbezogener Wechsel zu anderen Standorten eines Unternehmens kann nötig sein. Zusammen mit weiteren Prinzipien, wie der Einrichtung von Telearbeitsplätzen und mobilen Arbeitsplätzen kann die Forderung abgeleitet werden, daß die Pflege des Organisationsmodells „mitarbeiterspezifisch" von jedem Arbeitsplatz aus möglich sein sollte.

Dezentralität

Das Fehlen einer zentralen Einrichtung ist ein weiteres Merkmal einer VG. Ein IT-System zur Unterstützung einer solchen Gemeinschaft sollte deshalb auch keine solche zentrale Komponente benötigten. Vielmehr sollte es möglich sein, durch den Zusammenschluß der Aufbauorganisationen mehrerer autonomer Teilorganisationen eine virtuelle Gemeinschaft zu bilden. Das Top-level Design einer Teilorganisation wird in diesem Prozeß zum Bottom-level Design der virtuellen Gemeinschaft und darüber wird ein neues Top-level Design gebildet. Das Einrichten einer unabhängigen zentralen Komponente zur Verwaltung des gemeinsamen Organisationsmodells ist dabei aufgrund der möglichst kurzen Reaktionszeit schwierig zu realisieren. Wenn diese zentrale Komponente einem der Partner unterstellt ist, werden die anderen Partner von dieser Teilorganisation abhängig. Eine Lösung dieses Problems ist die gleichberechtigte dezentrale Verwaltung des Organisationsmodells in sich synchronisierenden Kopien (Repliken). In Kapitel II.II.4 wird der Lösungsansatz des GroupOrga Projekts mit Hilfe einer zugrundeliegenden Groupware-Plattform beschrieben.

Sicherheit

Zur Verwaltung von Aufbauorganisationsinformationen sind verschiedene Sicherheitsaspekte zu beachten, deren Lösungsansätze im Kapitel II.II.4 näher beschrieben werden. Hier soll nur beispielhaft ein Aspekt herausgestellt werden, der speziell für virtuelle Gemeinschaften gilt: Jede der Teilorganisationen veröffentlicht nur soviel Information, wie für die Kooperation mit den anderen Beteiligten benötigt. Der Grad des Zugriffs von Dritten auf die Aufbauorganisation der neu geschaffenen virtuellen Organisation und der Teilorganisationen muß nuanciert gesteuert werden können. In einem typischen Szenario wird der Zugriff von außen, falls überhaupt möglich, nur den *informierenden Zugriff* zur Suche von bestimmten Entitäten in einem Unternehmen umfassen.

II.3 Lösungsansätze mit Java als Technologie

Dieses Kapitel vergleicht die Lösung von Programmieraufgaben wie sie bisher erfolgte mit aktuellen *Java Programming* Technologien und stellt weiterhin heraus, wie der Einsatz von

Java helfen kann, Software-Systeme zu gestalten, die einige der im vorhergehenden Kapitel beschriebenen Anforderungen erfüllen.

Das nächste Kapitel verdeutlicht, daß der gemeinsame Einsatz von Java in Verbindung mit einer Groupware-Plattform zu einer Hebelwirkung führt, mit der weitere der dargestellten Anforderungen realisiert werden können.

Die herausragende Eigenschaft der Technologie Java ist die Auslegung als plattformunabhängige Programmiersprache[4]. In der bisherigen Umsetzung des GroupOrga Konzepts wurde bereits ein plattformunabhängiges Software-System erstellt. Bevor die Programmiersprache Java verfügbar war, wurde eine Programmiersprache verwendet, für die auf verschiedenen Plattformen und Betriebsystemen Compiler existieren. Für den GroupOrga Organization Modeler wird dazu die objektorientierte Programmiersprache C++ eingesetzt. Da der GroupOrga Organization Modeler eine graphische Benutzungsschnittstelle bereitstellen sollte, wird zudem eine Klassenbibliothek[5] eingesetzt, die ebenfalls auf mehreren Plattformen unterstützt wird. So konnte eine Software auf einer beliebigen Plattform erstellt werden. Um eine Version für ein anderes Rechnersystem, das auch unterstützt werden soll, zu erstellen, muß ein Compiler für die jeweilige Arbeitsumgebung bereitgestellt werden, mit dem der gesamte Quellcode einer bestimmten Version übersetzt wird (vgl. [13]).

Dieses Verfahren, Plattformunabhängigkeit zu erreichen, hat entscheidende Nachteile:

- Auf allen unterstützten Systemumgebungen muß jede Version neu kompiliert werden. Dazu muß im Entwicklerteam für jede der Plattformen die entsprechende Hardware, ein Compiler und das technische Wissen vorhanden sein, um eine solche Übersetzung des Quellcodes durchzuführen.
- Die Software muß in jeweils einer Version für jede unterstützte Plattform bereitgestellt werden und plattformspezifisch verteilt werden, d. h. den Benutzern einer bestimmten Umgebung müssen über einen Distributionskanal die Versionen für die entsprechende Plattform zukommen.

Zusammenfassend kann gesagt werden, daß für jede Hardware-Plattform, jede Betriebssystemversion und jede Programmversion eine neue Übersetzung des Quellcodes durchgeführt werden muß. Der gesamte Aufwand entfällt durch die Verwendung von Java als Programmiersprache mit der dazugehörigen Technologie. Mittels des Java Development Kit wird der Quellcode in Bytecode übersetzt und ist dann mit Hilfe einer Java Virtual Machine (JVM) auf jeder Plattform, die diese bereitstellt, ablauffähig. Aus der Präsenz der JVM ergibt sich eine Umkehrung der bisherigen Gegebenheiten: Die Systemumgebung wird an eine bestehende, für alle Systeme gleiche Software angepaßt, statt ein Softwareprodukt auf jede einzelne Systemumgebung abzustimmen. Somit sind keine plattformspezifischen Vertriebskanäle nötig, sowie kein spezielles Know-how für jede Plattform. Jede Version der Software muß nur einmal übersetzt werden. Falls eine neue Plattform oder ein neues Betriebssystem entsteht oder eine neue Version eines Betriebssystems erscheint, können

---

[4] "Write Once, Run Anywhere"

[5] StarView der Firma Star Division

einmal in Bytecode übersetzte Java-Programme darauf ausgeführt werden; so betrachtet sind Java-Programme zukunftskompatibel.

In Kapitel II.II.2 wurde ausgeführt, daß Organisationsmodellierung für jedes Mitglied eines Unternehmens ermöglicht werden soll. Daher gibt es, besonders für die Benutzer der Kategorien *informierender Zugriff* und *sporadische Strukturanpassungen* der Tab. 1, zusätzlich zu der zuvor geschilderten Plattformunabhängigkeit noch ein weiteres Argument für den Einsatz von Java: Es ist möglich, Java-Programme in der Form von Java-Applets in WebBrowsern ablaufen zu lassen. Im Bereich von Endbenutzeranwendungen bildet das Internet in Verbindung mit WebBrowsern als Front-End-Werkzeugen die weitaus größte Basis, um darauf aufbauend Anwendungen zu entwickeln, mit denen eine möglichst große und vielschichtige Gruppe von Benutzern erreicht werden soll.

Das verdeutlichte Ziel, *kurze Rüstzeiten* zu erreichen, soll hier nun näher mit einem Lösungsansatz adressiert werden. Eine entsprechende Installation auf einem WWW-Server vorausgesetzt, ist der initiale Aufwand für Anwender, die Java-Software zur Organisationsmodellierung zu nutzen, äußerst gering. Es muß lediglich eine URL angegeben werden, die sogar ggf. schon über einen Hyperlink von einer zentralen WWW-Seite (z. B. Homepage) des Unternehmens zu erreichen ist. Die entsprechende HTML-Seite wird dann zusammen mit dem Applet geladen und die Software kann sofort eingesetzt werden. In [16] wurde für Organisationsmodellierung im Team gefordert, daß jedes Organisationsmitglied die Daten bzgl. seiner Fertigkeiten und seines Know-hows pflegt, da er oder sie am besten die Entwicklung der eigenen Kompetenzen verfolgt und beurteilen kann. Mit dem in der vorigen Aussage beschriebenen Grundstock aus WebBrowser und Internet-Anschluß, unterstützt durch Java-basierten Organisationmodellierungswerkzeuge, kann diese Forderung erfüllt werden.

Eines der Ziele des GroupOrga Projekts ist, wie in Kapitel I.I.1 dargelegt, mittels IT korrespondierend mit einer Neuordnung von Strukturen eines Unternehmens die Wettbewerbssituation von Unternehmen zu verbessern. Die Java-basierten Werkzeuge zur Organisationsmodellierung schöpfen zusätzlich ein Rationalisierungspotential auf einer weiteren Ebene aus: Die Kosten für die Installation von Software reduzieren sich gegenüber den bisher eingesetzten Techniken drastisch. Propietäre Software, die unternehmensweit eingesetzt wird, muß auf jedem Rechner des Unternehmens installiert werden. Gleiches gilt für die Wartung von Software. Bei herkömmlichen Techniken müssen viele Änderungen (neue Anforderungen, technischer Wandel, Fehlerkorrekturen usw.) zusammengenommen werden, bevor dem Anwender eine neue Version zur Verfügung gestellt werden kann. Mit Java- basierten Softwareprodukten kann dem Anwender jede Änderung, die Software-Entwickler machen, sofort bereitgestellt werden.

Im vorigen Kapitel wurde gefordert, den Schulungsaufwand für Benutzer durch entsprechende Gestaltung der Software möglichst zu minimieren. Auch die mentale Belastung von Anwendern kann reduziert werden, da bestehendes Wissen aus mehreren Bereichen zu diesen Zwecken wieder angewendet werden kann, dazu gibt es in aktuellen Entwicklungen des GroupOrga Projekts zwei Ansatzpunkte. In den letzten Jahren ist durch die weite Verbreitung und intensive Nutzung des Internet bei vielen Anwendern von IT ein Basiswissen zur Benutzung von WebBrowsern und darin enthaltenen Informationen entstanden. Mit dem

Einsatz von Software-Werkzeugen, die in einem WebBrowser ausgeführt werden, kann dieses Wissen genutzt werden. Organisatorische Informationen können im Web allerdings in textueller Form auch ohne den Einsatz von Java bereitgestellt werden. Deswegen wird im zweiten Ansatzpunkt versucht, die Darstellung organisatorischer Strukturen möglichst intuitiv zu gestalten. Organigramme werden schon seit sehr langer Zeit in Unternehmen zur Darstellung von Organisationsstrukturen verwendet, die darin verwendete Darstellungsform ist vielen Organisationmitgliedern bekannt. Dieses Wissen kann genutzt werden, indem in Form einer Modellweltmetapher ähnliche Strukturen graphisch dargestellt werden. Zusätzlich wird versucht, eine ähnlich intuitive Darstellung von Workgroups zu finden.

Zu den erörterten Vorteilen, die sich durch den Einsatz der Technologie Java für das Projekt GroupOrga ergeben, kommt hinzu, daß die Programmiersprache Java eleganter und leichter verständlich ist, als zuvor weit verbreitete objektorientierte Programmiersprachen, wie z. B. C++. Unter anderem abstrahiert Java weiter von technischen Details als C++. Der teamorientierten Arbeit kommt entgegen, daß Java einige in die Programmiersprache integrierte Konstrukte enthält, die Software-Entwickler zu einer objektorientierten Stukturierung veranlassen.

Die immensen Vorteile, die durch Einsatz von Java-basierter Software entstehen, bergen auch Nachteile in sich, die im folgenden kurz beschrieben werden sollen. Im Projekt GroupOrga werden daher zunächst die Teile der Konzeption in Java umgesetzt, in denen die Pluspunkte von Java voll ausgenutzt werden und die derzeit in der Sprache noch vorhandenen Schwachstellen überlagern.

- Die Ausgereiftheit der Sprache Java und der dazugehörigen Technologien (JVM, JIT-Compiler) ist inzwischen sehr weit fortgeschritten. Bis zum heutigen Zeitpunkt ist allerdings die Verbreitung dieser erweiterten Java-Technologien in WebBrowsern noch nicht ausreichend ausgeprägt.
- Die Geschwindigkeit der Ausführung von Java Programmen kann, bedingt durch das zugrundeliegende Konzept der Plattformunabhängigkeit und der damit nötigen interpretativen Abarbeitung von Bytecode, nicht ganz an die von herkömmlichen Programmiersprachen heranreichen. Durch ausgeklügelte Konzepte (Just-in-Time-Compiler) werden aber Laufzeiten erreicht, die näher an die von anderen Programmiersprachen heranreichen.
- Die Applet-Sicherheit ist zum Datenschutz und Schutz der Computersysteme von Anwendern beim Benutzen von Java-Software notwendig, dennoch wird dadurch die Programmierung erschwert und einige Grenzen gesetzt, z.B. Einsatz von API-Zugriffen auf lokale Datenbanken.
- Bei jedem Start muß das Java-Programm neu von einem Server heruntergeladen werden. Dies kann bei starker Auslastung oder geringer Bandbreite des Netzwerks dazu führen, daß Benutzer eine lange Zeit warten müssen, bis eine Software einsatzbereit ist. Außerdem ist die Betriebsbereitschaft des Netzwerks Voraussetzung für den Einsatz der Java-Software; es kann nicht ohne Netzwerk gearbeitet werden, falls Java-*Applets* eingesetzt werden.

Einige der Nachteile werden durch die Weiterentwicklung der Java Technologie aufgelöst werden. Die beiden zuletzt genannten Nachteile können jedoch durch die Kombination von

Java-Programmen mit einer Groupwareplattform ausgeglichen werden. Außerdem entstehen aus dieser Kombination noch weitere Vorteile, die im nächsten Kapitel näher betrachtet werden.

II.4 Java und Groupware - Synergien für das Organisationsdesign

Ein entscheidender Grund im GroupOrga Projekt für die Integration von Organisationsdatenbanken in ein Groupwaresystem war, daß die Workflow- und Groupware-Anwendungen, die mit organisatorischen Informationen unterstützt werden sollen ebenfalls oft auf Groupwaresystemen basieren. Damit können die Anwendungssysteme direkt auf strukturelle Organisationsinformation zugreifen. Während dieser Grund weiterhin gültig ist und im Projektverlauf parallel weiterverfolgt wird, soll im folgenden eine Erweiterung - und nicht ein Ersatz - für diese Konzeption aufgezeigt werden.

Im intraorganisationalen Umfeld ist ein besonderer Vorteil dieser Groupware-basierten Architektur (vgl. Abb. 3), daß die Verteilung der Daten in den lokalen Netzwerkverbund von Workgroups oder Abteilungen mit Hilfe von Replikationsmechanismen unterstützt werden kann. Damit kann es für den größten Teil der Benutzer vermieden werden, daß Informationen, hier insbesondere Applets, über Weitverkehrsnetze (Wide Area Networks, WAN) zum Rechner des Anwenders transferiert werden müssen. Statt dessen können Anwender auf die Daten im lokalen Netzwerkverbund (Local Area Network, LAN) zugreifen, trotzdem können alle gemeinsam mit den Daten arbeiten. Die Nachteile von WAN Verbindungen sind im wesentlichen die oft sehr geringe Geschwindigkeit, die in Relation zu LAN Verbindungen geringe Zuverlässigkeit und evtl. die Kosten. Um dieses Ziel zu erreichen, kann auf einem Server in jedem lokalen Netzwerkumfeld eine Replik des Organisationsrepositorys abgelegt werden. Diese Repliken synchronisieren sich durch von der Groupwareplattform unterstützte Mechanismen. Vorteile der verteilten Datenhaltung nach diesem Prinzip sind, daß zum einen keine jederzeit verfügbaren WAN oder Internet-Verbindung notwendig ist. Des weiteren kann auf diese Daten sehr schnell zugegriffen werden, da sie sich im allgemeinen im Intranet-Umfeld befinden. Groupware-Datenbanken können mit der im Projekt GroupOrga eingesetzten Groupwareplattform[6] auch im WWW publiziert werden. Eine Java-Applikation kann dazu als Objekt in eine Datenbank integriert werden. Auf diese Weise wird die Rolle der Grouwareplattform, die unter anderem als Datenbankserver eingesetzt wurde, erweitert und kann jetzt zusätzlich auch als „Applikations"-Server (oder Appletserver) genutzt werden. Mit der Nutzung von Groupware als Plattform ist keine zentralisierte Komponente für das Organisationsinformationssystem nötig. Damit entspricht die technischen Realisierung der organisatorischen Konzeption von VG, die ebenfalls keine zentrale Einrichtung haben, was als Hinweis auf adäquates Systemdesign interpretiert werden kann.

II.5 Realisierung des Java-basierten GroupOrga Organization Modelers

In Abb. 4 ist der aktuelle Prototyp, des auf Java basierten GroupOrga Organization Modelers, der im GroupOrga Projekt entwickelt wird, gezeigt. Dabei wird bei der Entwicklung darauf

---

[6] Lotus Notes

geachtet, daß Teilergebnisse für die Bereiche *informierender Zugriff* und *sporadische Strukturanpassungen* der Skala in Tab. 1 genutzt werden können, bevor der GroupOrga Organization Modeler Funktionen bereitstellt, die für Nutzer der weiteren Teilbereiche der Skala benötigt werden.

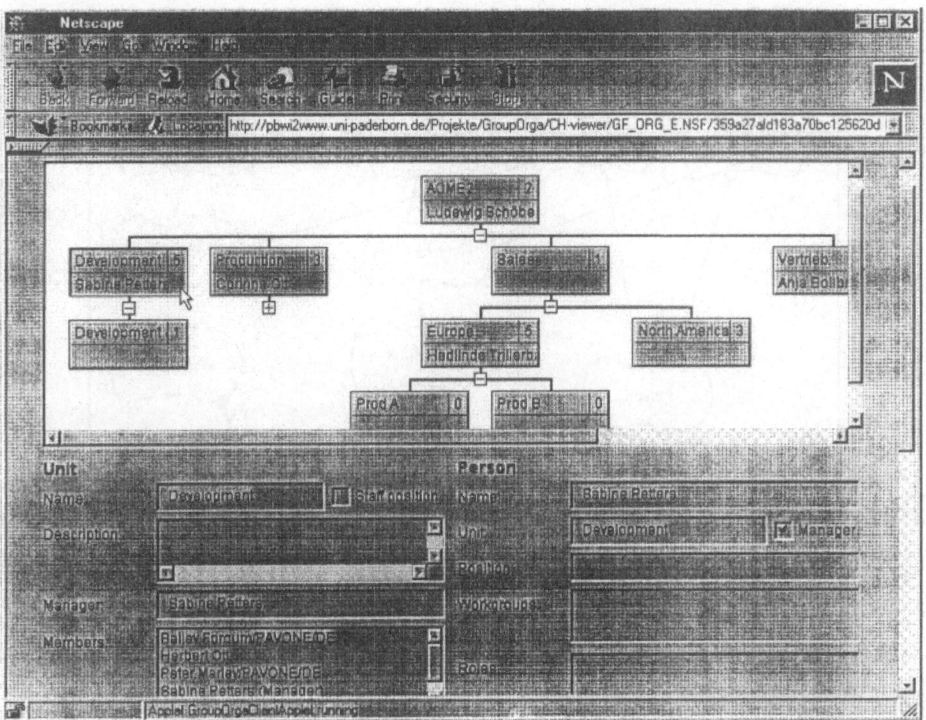

Abb. 4: Java basierter GroupOrga Organization Modeler

Abb. 5 zeigt ein abstraktes Klassenmodell des GroupOrga Organisation Modelers. Dieses enthält zunächst die Basisklasse *GroupOrgaModeler*. Die Klasse *accessOrgaDB* dient zum Austausch der Daten mit der Groupware-Datenbank. Die organisatorischen Daten werden aus der Groupware-Datenbank gelesen und zur hierarchischen Darstellung intern mit dem Modul *computeRelations* in eine Baumstruktur transformiert. Zudem gibt es zwei Klassen, die die verschiedenen graphischen Darstellungen der organisatorischen Informationen implementieren: *UnitGraph* und *WorkgroupGraph*. Diese setzen sich, von der unteren Ebene betrachtet aus einfachen Klassen zusammen, die dann zu komplexeren Klassen zusammengefügt und erweitert werden. Bspw. werden aus *Unit* schrittweise komplexere Klassen abgeleitet, die verschiedene Aspekte der Realisierung einer Abteilung beinhalten, *GUIUnit* enthält die Attribute und Methoden, die für die graphische Darstellung nötig sind. *HierarchicalUnit*, die Klasse, die auf *GUIUnit* aufbaut enthält die Bestandteile, die für die hierarchische Anordnung von Abteilungen benötigt werden. Auf Basis einer hierarchischen Abteilung wird dann im nächsten Schritt der Abteilungsgraph (*UnitGraph*) abgeleitet. Der

Bereich des Workgroup-Graphen wird duch ein etwas einfacheres, aber sonst ähnliches Klassenmodell repräsentiert.

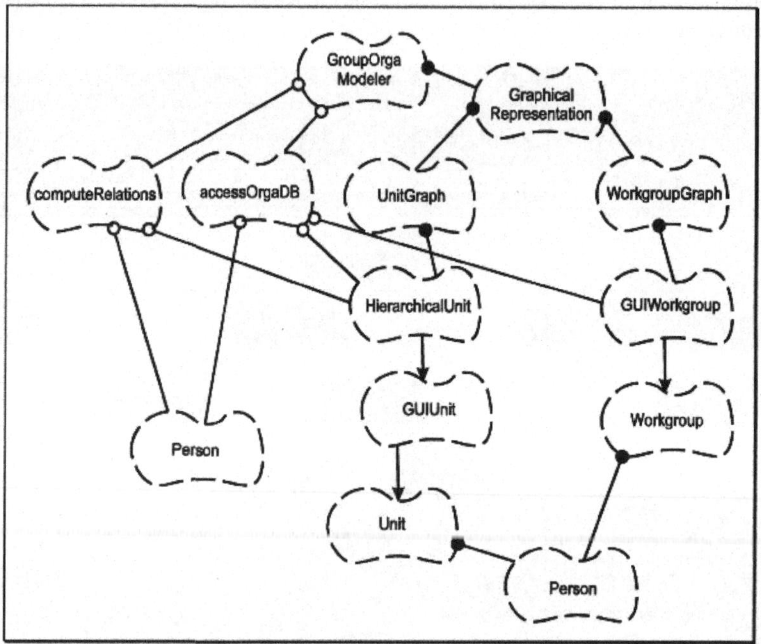

Abb. 5: Abstraktes Klassenmodell des GroupOrga Organisation Modelers

Das Phasenmodell der Verwendung des GroupOrga Organization Modelers umfaßt vier grundlegende Schritte. Darin wird zunächst im WebBrowser eine Anforderung an einen Server gesendet (1). Daraufhin wird das Applet auf den Client-Computer, auf dem der WebBrowser läuft, geladen (2) und dort gestartet (vgl. Abb. 6).

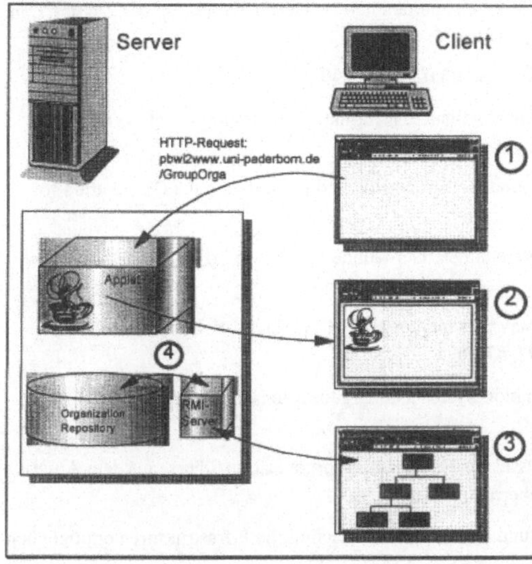

Abb. 6: Phasenmodell mit RMI

Nachdem das Applet gestartet ist, lädt es die Daten des Organisationmodells vom Server, wandelt sie in eine interne Repräsentation um und stellt sie graphisch dar. Danach besteht ein bidirektionaler Austausch zwischen dem Applet und dem Host, von dem es geladen wurde (3). Da die Applet-Sicherheit es nicht zuläßt, auf andere Ressourcen zuzugreifen, als auf den Server, von dem es geladen wurde, wird der Datenaustausch in der aktuellen Version mit Remote Method Invocation (RMI) durchgeführt. Dazu wird auf dem Groupware-WebServer zusätzlich ein RMI-Server installiert, der den Datenaustausch zwischen Groupware-Datenbank und Applet arrangiert. Der RMI-Server liest die Daten aus der auf dem Groupware-Server befindlichen Datenbank und sendet sie über das Netzwerk zu dem Applet. Im Applet veränderte Daten werden dann wiederum nach Übertragung zum RMI-Server von diesem wieder in die Groupware-Datenbank geschrieben.

II.6   Ausblick

Zum gegenwärtigen Zeitpunkt entfaltet das Java-Werkzeug seine besondere Stärke für *informierenden Zugriff* und *sporadische Strukturanpassungen* (siehe Tab. 1). Da die weitaus größte Anzahl von Nutzern in diese Kategorie gehören können die Vorteile von Java (vgl. Kapitel II.II.3) intensiv genutzt werden, während die Nachteile nicht zu vernachlässigen sind, aber von den Vorteilen überwogen werden.

Aufgrund der derzeit noch nicht erreichten Ausgereiftheit und der momentan noch nicht sehr hohen Geschwindigkeit von Java-Programmen ist eine sinnvoller Einsatz für *regelmäßige Änderungen* und *häufige, intensive Änderungen* derzeit noch nicht empfehlenswert. Für *intensive Nutzung* ist der Einsatz von proprietären Werkzeugen eher vertretbar, da die Anzahl der Installationen für diesen Bereich vergleichbar klein ist. Da die Nachteile von Java aber durch Fortentwicklungen in diesem Bereich der Softwaretechnologie ausgeräumt werden, werden in Java entwickelte Werkzeuge mehr und mehr Aufgaben unterstützen können, die im rechten Bereich des Organisationsdesignkontinuums angesiedelt sind (vgl. Tab. 1).

Literaturverzeichnis

[1]   M.S. Ackermann: Answer Garden: A Tool of Growing Organizational Memory, Ph.D. Thesis, MIT Sloan School of Management, 1994.

[2]   O. Arnold, W. Faisst, M. Härtling, P. Sieber: Virtuelle Unternehmen als Unternehmenstyp der Zukunft?, in: Handbuch der modernen Datenverabeitung, 32, 185, 1995, 8-23.

[3]   J. Byrne: The Virtual Corporation, in: Business Week, 08.02.1993, 1993, 37-41.

[4]   E.K. Clemons, M.C. Row: Rosenbluth International Alliance: Information Technology and the Global Virtual Corporation, in: IEEE, USA, 1/92, 1992, 678-685.

[5]   W.H. Davidow, M.S. Malone: The Virtual Corporation, Structuring and Revitalizing the Corporation for the 21st Century, Harper Business, NY, 1992.

[6]   W.H. Davidow, M.S. Malone: Das virtuelle Unternehmen, Der Kunde als Co-Produzent, Frankfurt - New York, 1993.

[7]   C.A. Gagne: Five years from now, groupware may be a memory, in: G.M. Hayes, K.I. MacLeish: Group(where), ComputerWorld, Nov., 24., 1997, 87-96.

[8]   V. Gurbaxani, S. Whang: The Impact of Information Systems on Organizations and Markets, in: Communications of the ACM, 34, 1991, 59-73.

[9]   J.R. Katzenbach, D.K. Smith: Teams: der Schlüssel zur Hochleistungsorganisation, übers. aus dem Amerik. "The Wisdom of Teams", Wirtschaftsverlag Ueberreuter, Wien, 1993.

[10] S. Klein: Virtuelle Organisation, Informations- und kommunikationstechnische Infrastrukturen ermöglichen neue Formen der Zusammenarbeit, in: WiSt, 23, 6, 1994, 309-311.

[11] P. Mertens: Virtuelle Unternehmen, in: Wirtschaftsinformatik, Vol. 36, No. 2, 1994, 169-172.

[12] P. Mertens, W. Faisst: Virtuelle Unternehmen - eine Organisationsstruktur für die Zukunft?, in: technologie & management, vol 44, 1995, no. 2, 1995, 61 -68.

[13] S. Meyer: Neue Konzepte zur computergestützten Organisationsmodellierung, Untersuchung und Konzeption eines computergestützten Hilfsmittels zur Organisationsmodellierung zum Einsatz in innovativen Workflow Management Systemen (Organization Object Modeler), Diplomarbeit, Universität Paderborn, Lehr- und Forschungseinheit Wirtschaftsinformatik 2, August, 1996.

[14] N.N.: Neue Technologien für virtuelle Organisationen. Groupware als Katalysator für das virtuelle Unternehmen, in: Gablers Magazin, 6-7, 1994, 28-30.

[15] M. Ott, L. Nastansky: Modelling Organizational Forms of Virtual Enterprises, The Use of CSCW Environments for a Team Based, Distributed Design of Virtual Organizations, in: J. Griese, P. Sieber (Eds.): VoNet, The Newsletter @ http://www.virtual-organization.net/news/main.html, Institute of Information Systems Department of Information Management University of Berne, Vol. 1, No. 4, September 1, 1997, 20-39.

[16] M. Ott, L. Nastansky: Groupware Technology for a new Approach to Organization Design Systems, To be published in: F. Nunamaker jr., R.H. Sprague (Eds.): Proceedings of the 31st Hawai'i International Conference On System Sciences, Organizational Systems and Technology, Kona, Hawaii, January 6-9, IEEE Computer Society Press, Los Alamitos, 1998.

[17] M. Ott: Groupware, Charakterisierung und technologische Perspektive, in: WiSt - Wirtschaftswissenschaftliches Studium, C.H. Beck, Vahlen, München, Frankfurt a.M., Februar, 1997, 90-94.

[18] P. Sieber: Annotated Bibliography zum Thema Virtuelle Unternehmen, Arbeitsbericht Nr. 73, Arbeitspapier der Reihe "Informations und Kommunikationssysteme als Gestaltungselement Virtueller Unternehmen", Bern Leipzig Nürnberg, 4, 1995.

[19] E.W. Stein: Organizational memory: Review of concepts and recommendations for management, in: International journal of Information Management, 15(2), 1995, 17-32.

[20] H. Wildemann: Ein Ansatz zur Steigerung der Reaktionsgeschwindigkeit von Unternehmen: Die Lernende Organisation, in: Zeitschrift für Betriebswirtschaftslehre, Ergänzungsheft, Gabler, Wiesbaden, 3, 1995, 1-23.

[21] H.A. Wüthrich, A. Philipp: Das Grenzenlose Unternehmen, Virtuelle Netzwerkstrukturen - organisatorischer Trend für das 21. Jahrhundert?, in: zfo, Schäffer Poeschel, Stuttgart, 6, 1996, 383-384.

[22] J. Yates: For the record: The embodiment of organizational memory, 1850-1920, Business and Economic History, 2nd series, Nineteen, 1990.

Adressen der Autoren

Dipl.-Wirt. Inform. Marcus Ott
Universität-GH Paderborn
Lehr-/Forschungseinheit Wirtschaftsinformatik 2
Warburger Str. 100,
D-33098 Paderborn
Tel. ++49-5251-603368
Email: Mott@notes.uni-paderborn.de
http://fb5www.uni-paderborn.de/winfo2/MarcusOtt

Carsten Huth
Universität-GH Paderborn
Lehr-/Forschungseinheit Wirtschaftsinformatik 2
Warburger Str. 100,
D-33098 Paderborn
Tel. ++49-5251-603368
Email: Chuth@notes.uni-paderborn.de
http://fb5www.uni-paderborn.de/winfo2/GroupOrga

Verschränkung von Perspektiven durch Aushandlung

Thomas Herrmann*, Gerry Stahl**

*Universität Dortmund, Informatik und Gesellschaft,

** University of Colorado, Boulder, Center of LifeLongLearning and Design

Abstract:

To foster collaborative learning a system should support perspectival viewing and negotiation in a natural way. A systeme - called WebGuide - is described which supports teams of students conducting collaborative research on the web; it lets them share bookmarks, queries, notes and summaries that arise in their research. The shared information space is viewed through various perspectives so that all participants can construct their own personal view of the materials, rearranging them under different categories and generating new summaries or annotations. All information can be modified within a personal perspective. Participants can propose that items they have created be incorporated in the team's shared perspective. For a proposed item to be accepted in the team perspective, team members must deliberate and agree to accept it. Therefore, a negotiation mechanism is provided.

1 Einleitung

1.1 Zielsetzung

Das WWW ist ein naheliegendes Instrument zur Unterstützung von kooperativer Arbeit. Insbesondere stellt es eine vielfältige und entsprechend unüberschaubare Informationsbasis dar, auf die sich kooperative Arbeit beziehen kann. In Kooperationsbeziehungen hat jeder einzelne Akteur seine eigene Perspektive, unter der er diese Informationsbasis nutzt. Mit Perspektive ist hier gemeint, daß ein eingeschränkter Ausschnitt der Informationsbasis betrachtet wird, der festgehalten, kategorisiert, kommentiert und modifiziert werden kann. Im Rahmen kooperativer Arbeit ist es erforderlich, daß diese Perspektiven miteinander verschränkt werden. Das kann bedeuten, daß man die Perspektiven anderer zur Kenntnis nimmt, daß man einen Teil von anderen Perspektiven in die eigene übernimmt etc. Eine besondere Herausforderung stellt es dar, verschiedene Perspektiven zu einer gemeinsamen Perspektive zusammenzufügen. Im Falle einer wissenschaftlichen Arbeitsgruppe oder auch für ein aus Schülern oder Studenten gebildetes Team können wir uns vorstellen, daß zunächt in Einzelarbeit zu einem Thema Informationen mit Hilfe des WWW gesammelt und bearbeitet werden. Zu einem gegebenen Zeitpunkt wird es darauf ankommen, diese Informationen miteinander zu verbinden, das heißt zu entscheiden, welche Informationen dem weiteren Arbeiten des gesamten Teams als gemeinsame Grundlagen dienen soll. Ein solcher Entscheidungsprozeß bedarf in der Regel mehrerer Verständigungs- und Aushandlungsschritte. Ein Teil dieser Schritte kann durch geeignete Groupware-Konzepte selbst unterstützt werden. Sofern sich die Aushandlung auf Perspektiven bezieht, die mit dem WWW verbunden sind, ist es sinnvoll, die Unterstützung der Aushandlung ebenfalls auf das WWW aufzubauen.

Wir schlagen im folgenden ein Konzept vor, mit dem man auf Basis des WorldWideWeb Informationssammlungen einzelner Mitglieder eines Teams miteinander verschränken kann und sowohl inidividuell als auch kollektiv bearbeiten kann.

1.2 Eigene Vorarbeiten

Wir kombinieren zwei Ansätze: Zum einen Collaboration with Perspectives von Stahl (1993, Kapitel 9) und zum anderen Aushandelbarkeit von Herrmann (1994, 1995).

Etwas vereinfacht lassen sich die wesentlichen Eigenschaften des Perspektiven-Ansatzes von Stahl wie folgt beschreiben:

1. Einzelne Teammitglieder verfügen jeweils über eine eigene Informationsbasis, Perspektive genannt, die aus einzelnen Informationseinheiten besteht, welche mittels einer Hyperstruktur miteinander verknüpft sind. Die Informationseinheiten können also als Knoten aufgefaßt werden, die durch gerichtete Kanten, hier Links genannt, miteinander verbunden werden und ein Netz aufbauen. Die Richtung kann im Sinne einer Vererbung verstanden werden. Perspektiven sind durch Sub-Netze repräsentiert.

2. Teammitglied A(ndrea) kann den Inhalt eines Knotens X der Perspektive von B(ert) in ihrer eigenen Perspektive mittels eines Hyperlinks zugänglich machen. Wenn B den Inhalt des Knotens ändert, ändert sich auch der für A zugängliche Inhalt. Wenn A den Inhalt des in ihrer Perspektive dargestellten Knotens ändert, wird der Link zu B unterbrochen und ein neuer Knoten mit einem neuen Link zu Andreas Perspektive erzeugt. A kann also den Inhalt von X ändern, löschen, umstellen oder auch in X enthaltenen Links umhängen, ohne daß sich in Bs Perspektive etwas ändert. Ob sich in der Perspektive anderer Teammitglieder etwas ändert, hängt davon ab, ob sie X von A oder von B geerbt haben.

3. Teammitglied A(ndrea) kann den Inhalt eines Knotens X der Perspektive von B(ert) in ihre eigene Perspektive direkt kopieren. Es besteht kein Link zwischen den Perspektiven, sondern nur ein Link, der den neuen Knoten X' zu As Perpsektive zuordnet.

4. A kann entscheiden, ob sie mit einem Link zu einem Knoten auch das dazugehörige Sub-Netz übernimmt oder ob sie ein eigenes Sub-Netz aufbauen will.

5. Knoten und Links können editiert und insbesondere auch gelöscht werden.

6. Die Möglichkeit der Vererbung zwischen Perpektiven erlaubt es einer Gruppe, die Verschränkung ihres Wissens zu strukturieren. Aus zwei oder mehreren individuellen Perspektiven lassen sich (Sub-)Team-Perspektiven bilden. Es entsteht eine Hierarchie von Perspektiven, die sich in einem Baum repräsentieren lassen. Dieser Baum stellt dann die Beziehungen des Informationsaustauschs zwischen Individuen bzw. zwischen (Sub-)Teams mittels einer Vererbungshierarchie dar. Ein solcher Baum kann dann so ausgewertet werden, daß die untergeordnete Perspektive immer automatisch die Knoten der übergeordneten Perspektive erbt.

Es ist ein wesentlicher Vorteil diese Konzeptes, daß Teammitglieder Inhalte der Teamperspektive erben können, ohne sie neu generieren zu müssen und daß sie mit diesen Inhalten experimentieren können, ohne die Sicht der anderen Teammitglieder auf die

Teamperspektive dadurch zu beeinträchtigen. Dieser Vorteil ist solange wirksam, wie Teammitglieder die Perspektiven der anderen nur zur Anreicherung der eigenen Perspektive nutzen wollen. Sobald es aber für ein Teammitglied von Interesse ist, welche Inhalte in der Perspektive eines anderen Mitglieds enthalten sind, stößt das Konzept an Grenzen, da es nicht möglich ist, die Inhalte anderer zu ändern oder zu fixieren. Für kooperatives Arbeiten ist es jedoch ausschlaggebend, daß sich die zugrundeliegenden Perspektiven zumindest partiell überlappen, um eine erfolgreiche Verständigung und Koordination zu erreichen (s. z.B. [4]). Die zugrundeliegenden subjektiven Auffassungen werden zu Gunsten einer Intersubjektivität verschränkt ([9], S. 28). Im Kontext der Informatik und Software-Entwicklung wurde an verschiedenen Stellen verdeutlicht, daß die beteiligten Entwickler und Nutzer ein System immer unter verschiedenen Perspektiven betrachten und beurteilen und daß diese Unterschiede zu berücksichtigen sind ([7], S. 91).

Das Konzept systemgestützter Aushandlung sieht vor, daß man Veränderungen von Systemeigenschaften oder von Informationen auch dann vornehmen kann, wenn sie potentiell mit den Interessen anderer kollidieren. Eine solche Änderung wird zunächst vorgeschlagen. Mit Hilfe des gleichen Systems, mit dem diese Änderung vorbereitet und durchgeführt wird, werden auch die von dieser Änderung Betroffenen informiert und in die Lage versetzt, auf den Vorschlag zu reagieren. Folgende Reaktionsmöglichkeiten sind nach Herrmann [11] vorgesehen:

1. der Vorschlag wird akzeptiert,

2. der Vorschlag wird auf Widerruf akzeptiert,

3. der Vorschlag wird abgelehnt,

4. Modifikation des Vorschlages,

5. Abbruch der systemgestützten Verhandlung mit dem Vorschlag, die Kommunikation auf direkterem Wege zu führen (z.B. face-to-face oder telefonisch),

6. Jede der vorangegangenen Reaktionsmöglichkeiten kann zusätzlich kommentiert werden.

Dieses Konzept wurde für Situationen entwickelt, in denen zwei Systembenutzer darüber verhandeln, ob eine Systemanpassung erfolgen darf oder nicht. Dieser Ansatz hat den Zweck, Individualisierbarkeit und Steuerbarkeit (im Sinne von ISO 9241, Teil 10) auch für Groupware umsetzbar zu machen. Die Aushandlung kann in mehreren Zyklen stattfinden, indem der Vorschlagende wiederum auf die Reaktion des Betroffenen reagiert. Es sind Aushandlungsregelungen festzulegen, die bestimmen, wieviele Aushandlungszyklen stattfinden können, wieviel Zeit bis zu einer Reaktion verstreichen kann, was passiert, wenn das Zeitlimit überschritten ist etc. Zweck dieses Aushandlungskonzeptes ist es, möglichst schnell die Routinefälle von Ablehnung, Zustimmung oder einfachen Modifikationen der Vorschläge zu erledigen und gezielt herauszufinden, zu welchen Vorschlägen ein intensiverer Kommunikationsprozeß stattfinden muß. Auf diese Art und Weise erhält man *eine* gemeinsame Ausgangsbasis, auf der Kooperation aufbauen kann. Der Nachteil besteht darin, daß dieses Konzept nur für zwei Aushandelnde entworfen ist und daß bei der Erweiterung für mehrere Teilnehmer auch mehr Zeit verstreichen kann, bis es zu einer Einigung auf eine gemeinsame Ausgangsbasis kommt. Das ursprüngliche Aushandlungskonzept [11] sieht vor, daß solange nicht mit der geänderten Version gearbeitet werden kann, bis der

Aushandlungsprozeß abgeschlossen ist. Das mag für Veränderungen der Systemfunktionalität sinnvoll sein, nicht aber aber für Veränderungen von Informationsbasen. Für diese bietet das Konzept der Verschränkung verschiedener, individueller und gemeinsamer Perspektiven den Vorteil, daß man bis zur Erzielung des Aushandlungergebnisses mit der individuellen Perspektive weiterarbeiten kann.

1.3   Vergleich mit anderen Ansätzen

Ein wesentlicher Mechanismus zur Unterstützung kooperativer Arbeit mit gemeinsamen Material sind **Hypertext- und Hypermediastrukuren**, wie sie letztlich auch zumindest partiell mit dem WWW realisiert sind, wobei nach wie vor Erweiterungsanforderungen bestehen [3]. Wie oben schon angedeutet, basiert das Konzept verschränkter Perspektiven auf Knoten und Links. Der Ansatz von Stahl [18] geht jedoch hinsichtlich der Verwaltung und Manipulierbarkeit von Links über das hinaus, was in der Regel mit Hypertextstrukturen angeboten wird. Besonders zu erwähnen ist die Möglichkeit, Knoten zu editieren, die man mittels eines Links innerhalb der eigenen Perspektive darstellt. Da es aus Konsistenzgründen nicht möglich ist, auch geerbte Knoten zu ändern, muß eine Kopie des Knotens erstellt werden, der Link wird abgehängt und die Kopie kann dann bearbeitet werden. Solche Konzepte sind zum Beispiel bei **Betriebssystemen** bereits realisiert [6] ohne jedoch in eine umfassendere Umgebung zur Unterstützung kooperativer Arbeit eingebettet zu sein.

Ein anderes Konzept, das unmittelbar Assoziationen zu der beschriebenen Differenzierung verschiedener Perspektiven anregt, ist die Ermöglichung verschiedener **Views auf Datenbanken.** In Abhängigkeit von der Struktur einer Datenbank kann bekanntlich festgelegt werden, daß verschiedene Benutzer nur bestimmte Ausschnitte zur Kenntnis nehmen können oder dürfen. Durch Selektionsmechanismen (etwa bzgl. der Reihen einer Tabelle, die die Daten repräsentiert) oder Views (bzgl. der Spalten der Tabelle) können verschiedene Perspektiven definiert werden. Der Unterschied zu dem hier verfolgten Ansatz besteht darin, daß Benutzer ihre eigene Perspektive selbst definieren können und zwar unter Bezugnahme auf andere individuelle oder gemeinsame Views und Selektionen. Es ist nicht auszuschließen, daß man mit dem View-Konzept bei Datenbanken ähnliche Mechanismen konfigurieren kann, allerdings ist dies nicht das hauptsächliche Anliegen dieses Konzeptes. Vielmehr wird es dabei immer   notwendig sein, auf andere Funktionen von Datenbanken, wie etwa Anfragemechanismen   zuzugreifen. Insofern kann man sagen, daß Datenbanken den beschriebenen Perspektivenansatz nicht per se beinhalten, sehr wohl aber unter Ausnutzung ihrer gesamten Funktionalität genutzt werden können, um ihn in Verbindung mit dem WWW zu realisieren.

Unter **Organizational Memories** verstehen wir grob gesehen einen Ansatz, bei dem eine strukturierte elektronische Ablage für alle Arten von Dokumenten geschaffen wird, die im Rahmen computergestützter Kooperation und Kommunikation unter den Teilnehmern ausgetauscht werden (s. zum Beispiel [1]; [14]). Es stellt dabei eine wesentliche Anforderung dar, diese elektronische Ablage parallel zu der Kooperation aufzubauen, ohne daß die Kooperierenden hierdurch gestört werden oder zusätzliche Arbeit notwendig wird und daß dennoch eine Struktur entsteht, die es ermöglicht, später möglichst direkt die relevanten Inhalte zu finden. In diesem Sinne ist das Konzept verschränkter Perspektiven nicht zum Aufbau von Organizational Memories gedacht, da es zusätzlicher Arbeits- und

Entscheidungsschritte bedarf, diese Perspektiven aufzubauen, wodurch wiederum die eigentliche Kooperation gestört werden könnte. Vielmehr ist der Aufbau solcher Perspektiven sinnvoll, um ein vorhandenes Organizational Memory mit Hinblick auf eine neue kooperative Aufgabe auszuwerten, indem man Teile des Memories in die Perspektiven aufnimmt, die zur Bearbeitung der neuen Aufgabe aufgebaut und ausgehandelt werden. Perspektivität erlaubt es, Organizational Memories in verschiedener Weise zu organisieren und zu kategorisieren. Auf diese Weise können schnell neue Strukturen von Informationen geschaffen werden, die an die Erfordernisse der jeweiligen kooperativen Aktivitäten angepaßt sein können.

Ein weiteres Konzept, das einen Raum vorhandener Informationen in Abhängigkeit von den Präferenzen der Individuen oder Teams ordnet, ist **Group Lens** [16]. Hier wird mittels statistischer Auswertungen automatisch festgestellt, welche Mitglieder einer Gruppe sich für ähnliche Themen interessieren. Solche Korrelationen werden genutzt, um Informationen, die von einem Gruppenmitglied als interessant angesehen werden, an diejenigen anderen Gruppenmitglieder weiterzuleiten, mit denen eine Korrelation hinsichtlich der Themen festgestellt wurde. Auf diese Weise wird also automatisch eine gemeinsame Perspektive auf ausgewählte Informationsausschnitte hergestellt. Dieser Ansatz ist insbesondere für neu auftretende Informationen hilfreich. Der Automatismus ist in dem Perspektivenkonzept nach Stahl [18] insofern enthalten, als man das Sub-Netz erben kann, das sich an einen Knoten anschließt. Auf diese Weise kann man auf alle Informationen Zugriff erhalten, die der Inhaber eines Knotens an selbigen mittels Links anhängt. Ansonsten unterstützt Group Lens nicht die aktive Selektion bestimmter Information, die auf einer bewußten Entscheidung der Akteure beruht.

Die bisher beschriebenen Ansätze enthalten keine systembasierten Aushandlungsmechanismen. Wulf [21] hat den Ansatz von Herrmann zum Zwecke des **Konfliktmanagements bei Groupware** fortentwickelt. Dabei unterscheidet er drei Möglichkeiten, mit denen man eine Beeinflussung der eigenen Interessenssphäre durch andere Groupwarenutzer vermeiden kann:

1. Der Benutzer kann unterbinden, daß ihn andere durch ihre Art der Systemnutzung in unerwünschter Weise beeinflussen;

2. der Benutzer kann den Einfluß anderer nicht unterbinden, aber nachvollziehen;

3. der Einfluß anderer ist aushandelbar.

Wir meinen dagegen, daß es den Nutzern immer möglich sein sollte, bei Bedarf wechselseitig aufeinander zu reagieren, und sei es nur mittels Kommentaren. Die wechselseitigen Reaktionen sollten idealer Weise immer mit Hilfe desselben Systems erfolgen können, auf das sich die Reaktionen inhaltlich beziehen.

Die deutlichsten Parallelen mit der systembasierten Aushandlung finden sich bei **Decision Support** und **Meeting Support Systemen** (eine Übersicht gibt [8]. Zum einen kann man dort auf Vorschläge, die von anderen unterbreitet wurden, modifizierend reagieren, indem man sie direkt durch eigene Vorschläge ergänzt. Zudem ist es möglich, Kommentare an die Vorschläge anzuhängen. Etwas elaboriertere Systeme, die man als Nachfolger des ARGNOTER-Konzepts ([19], S. 345ff.) ansehen kann, erlauben es, Kommentare zu klassifizieren, je nach dem, ob sie als Pro- oder Gegenargument zu einem Vorschlag zu verstehen sind. Diese Argumente stellen eine Parallele zu Ablehnungs- oder

Zustimmungsvoten dar, werden aber nicht immer als solche automatisch ausgewertet. Zur Berechnung von Zustimmung oder Ablehnung ermöglichen es einige Systeme, daß die Teilnehmer Bewertungen für die Vorschläge abgeben, indem sie z.B. eine vorgegebene Anzahl von Punkten über die Vorschläge verteilen. Die Vorschläge, die die meisten Punkte erhalten, gelten dann als diejenigen, auf die sich eine Gruppe geeinigt hat. Um solche Abstimmungen gerecht zu organisieren, sind verschiedene Randbedingungen zu beachten; einen guten Überblick und Lösungsdiskussionen hierzu gibt Ephrati et. al. [5]. Eine Anwendung bei Terminkalendern beschreiben Sen et. al. [17]. Mit solchen Abstimmungsverfahren wird häufig eine große Anzahl von Perspektiven abgearbeitet, die dann meistens nicht in angemessener Weise inhaltlich gewürdigt werden, insbesondere, wenn keine Kommentierung oder Diskussion zu den einzelnen Punkten erfolgt. Das hier vorzustellende Konzept von Perspektivenverschränkung durch Aushandlung versucht, beides miteinander zu verbinden.

Ansätze zur Aushandlung finden sich zum Teil auch in Konzepten, die auf **Agenten** aufbauen. Martial [15] schlägt ein solches Konzept zur Lösung von Konflikten bei der Ressourcenzuweisung vor. Neben den oben genannten Reaktionsmöglichkeiten im Rahmen von Aushandlungsprozessen differenziert er zusätzlich zwischen Modifizierung eines Vorschlags und Unterbreitung eines Gegenvorschlags, während wir versuchen, ohne diese Unterscheidung auszukommen. Bayer [2] entwickelt ein Konzept, bei dem ein Teil der Aushandlungsarbeit von Agenten übernommen wird, um die Benutzer in Standardsituationen zu entlasten.

Es zeigt sich, daß die beschriebenen Konzepte jeweils nur Teilaspekte des hier darzustellenden Lösungskonzeptes beinhalten. Andererseits beinhalten sie auch wertvolle Anregungen, die bei der Ausgestaltung von Möglichkeiten für die Perspektivenverschränkung durch Aushandlung zu berücksichtigen sind.

2 Beispiele für die Notwendigkeit verschränkter Perspektiven

Um die Entwicklung eines konkreten Lösungskonzepts für die Verschränkung von Perspektiven vorzubereiten, analysieren wir einige Beispielfälle, aus denen sich entsprechende Anforderungen ableiten lassen. Wir gehen davon aus, daß das Lösungskonzept für jeden Anwendungsbereich unterschiedlich auszugestalten ist. Wir stellen hier dar, wie man ohne Unterstützung durch ein Computersystem verfahren muß, wenn verschiedene Sichtweisen aufeinander abgestimmt werden müssen. Hierdurch wollen wir die Probleme verdeutlichen, die es zu beheben gilt.

2.1 Verwaltung von Stichworten zur Klassifizierung von Einträgen in einer Literaturdatenbank

Wissenschaftliche Teams nutzen häufig eine gemeinsame Literaturdatenbank für ihre Arbeit. Ein Teil der Einträge in einer solchen Datenbank ist oft nur für einzelne Individuen interessant, ein anderer Teil hingegen ist für alle oder für Untergruppen relevant. Während die Basisangaben zu einer Literaturstelle meistens einheitlich sein sollten, kann es bei der Bewertung oder Kommentierung einer solchen Stelle zu unterschiedlichen Auffassungen kommen. Dieses Problem kann man lösen, indem für jeden Nutzer eine eigene Sicht definiert wird, die ihm solche Bewertungen und Kommentare zeigt, die er als seinen Vorstellungen entsprechend ausgewählt hat. Diese Individualisierung macht allerdings bei solchen Informationen, die die Zusammenarbeit unterstützen sollen, wenig Sinn. Noch deutlicher als bei den Kommentaren wird dies am Beispiel von Schlagwörtern, mit Hilfe derer man Literatureinträge kategorisiert.

Solche Schlagwörter haben nicht nur den Zweck, das eigene Gedächtnis zu unterstützen, sondern auch anderen das Auffinden von Literatur zu erleichtern. Deshalb sollte der Katalog von Schlagwörtern, die in einem Team zum Einsatz kommen, immer wieder aktualisiert und abgestimmt werden. Geht man von einer einfachen Liste aus, so können Schlagwörter etwa zusammengefaßt oder differenziert werden, sie können gelöscht oder durch neue ersetzt werden oder es können neue, zusätzliche Schlagwörter eingeführt werden. Wer immer eine solche Änderung vornimmt, sollte entscheiden, ob diese nur für ihn persönlich relevant ist, oder ob er der gesamten Gruppe vorschlägt, mit dem geänderten Schlagwort zu arbeiten. In letzerem Fall kann dann zum Beispiel der Chef entscheiden, ob der Vorschlag angenommen wird, oder man kann per Umlaufverfahren abstimmen. Beide Vorgehensweisen haben den Nachteil, daß es nicht zu einer inhaltlichen Diskussion über den Änderungsvorschlag kommt, das Umlaufverfahren nimmt zudem auch viel Zeit in Anspruch. Die inhaltliche Diskussion ist in der Regel wichtig, um das gemeinsame Verständnis eines Schlagworts zu erhöhen. Ein Ausweg wäre es, die Annahme eines Änderungsvorschlages im Rahmen einer Gruppendiskussion zu entscheiden. Dieses Verfahren hat den Nachteil, daß viel unnötige Diskussionszeit auf banale Entscheidungen verwendet wird.

Es geht darum, ein systemgestütztes Verfahren zu entwicklen, mit dem man schnell Kommentare austauschen kann, zu einer Abstimmung kommt oder feststellen kann, welche Änderungen einer intensiveren Diskussion bedürfen. An der Universität Dortmund wird zur Zeit ein WWW-basierter Prototyp für ein solches System entwickelt.

2.2 Fortentwicklung eines Glossars

Für die Arbeit in Teams kann es häufig von Nutzen sein, eine Liste der wesentlichen Grundbegriffe samt Definitionen aufzustellen, hierarchisch zu strukturieren und kontinuierlich zu pflegen. Änderungen von Text oder bezüglich der Einordnung in der Hierarchie bedürfen bei einem Glossar, das einer Gruppe als Kommunikationsbasis dient, der Zustimmung aller, ebenso die Eintragung neuer Begriffe. Es sollten Verfahren bereit stehen, mit denen einfache Änderungen, wie etwa das Beheben von Tippfehlern, möglichst unbürokratisch erledigt werden können. Das oben genannte Umlaufverfahren ist hierfür nicht geeignet, sonderen eher der Weg über eine zentrale Entscheidungsinstanz, die feststellt, ob eine Änderung einer intensiveren inhaltlichen Diskussion bedarf. Daneben sollte es Mitgliedern einer Gruppe möglich sein, ein individuelles Glossar aufzustellen, wenn es sich etwa um ein interdisziplinäres Wissenschaftlerteam handelt. In diesem Glossar kann festgehalten werden, zu welchen Begriffen einzelne Gruppenmitglieder eine abweichende Auffassung im Unterschied zu dem restlichen Team haben. Ein System verschränkter Perspektiven mittels Aushandlung sollte in der Lage sein, die hier skizzierten Vorgänge zu unterstützen und zu ergänzen.

2.3 Kooperatives Lernen in Schülergruppen mit asynchronen Aktivitätsphasen

Wir gehen hier von einem Fall aus, der den Hintergrund einer aktuellen Systementwicklung abgibt, die am Center of LifeLongLearning and Design der University of Colorado, Boulder unter dem Titel WebGuide verfolgt wird. Ein Lehrer gibt z.B. Schülern die Aufgabe, zum Thema „Historische Indianerkulturen" individuell unter Nutzung des WWW zu recherchieren und die Ergebnisse ihrer Recherche zusammenzufassen. Er läßt vier Untergruppen zu den Themen Inkas, Mayas, Azteken und Anasazi bilden. Die Schüler können dann Bookmarks sammeln und diese kommentieren oder kurze Texte aus den Quellen, die sie finden, zusammenstellen. Um das Ergebnis ihrer Recherche überschaubar zu halten, werden sie dazu angeregt, Kategorien zu bilden, mit denen sie ihre Einträge zusammenfassen und ordnen. Danach sollen sie ihre Ergebnisse in den einzelnen Untergruppen zusammenführen. Dies stellt in der Regel eine Überforderung dar, da einzelne, von den Schülern selbst gewählte Strukuren oftmals kaum kompatibel sind. Schüler sind unter Umständen gezwungen, einen Teil ihrer Arbeit wegzuwerfen, was demotivierend ist. Es wird unnötige Doppelarbeit geleistet. Es ist für Schüler schwierig, ihre Ergebnisse zusammenzufassen, wenn sie nicht angeleitet werden, die dazu notwendige Diskussion zu strukturieren.

Es ist ein System anzubieten, das das Zusammenführen der Arbeitsergebnisse schon sehr früh und kontinuierlich unterstützt und die dazu notwendigen Abstimmungprozesse strukturiert und erleichtert, so daß die Schüler Möglichkeiten aufgezeigt bekommen, wie man einen Konsensbildungsprozeß durchführen kann. Wir wählen dieses Beispiel, um unser Konzept der Verschränkung von Perspektiven durch Aushandlung zu erläutern. Zum einen weist das Beispiel auf eine hohe Komplexität hin, anhand derer sich die Lösungsansätze umfassend darstellen lassen. Zum anderen handelt es sich um ein Beispiel für Computer Supported Cooperative Learning, einer aktuellen Forschungsrichtung (s. [13]), die für innovative Nutzungen des WWW von besonderer Bedeutung ist.

3 Das Lösungskonzept

3.1 Voraussetzungen

Schüler sollen einzelne Teile der Arbeit anderer in ihre Perspektive übernehmen können und sie sollen frühzeitig Vorschläge unterbreiten können, welche Teile ihrer eigenen Perspektive oder der Perspektive anderer in die Team-Perspektive eingehen sollen, die letztlich die Zusammenfassung der Arbeit repräsentiert. Diese Anforderung setzt es voraus, daß Informationen in kleinen Portionen dargestellt und gesammelt werden können. Letztlich kann man nur mit Bezug auf kleine, überschaubare Informationseinheiten aushandeln, welche übernommen, geändert oder gelöscht werden sollen. Die Abbildung 1 stellt eine geschachtelte Datenstruktur der Informationseinheiten dar, die für WebGuide relevant sind.

Eine entscheidende Voraussetzung von WebGuide ist es, daß Schülern Webseiten im Sinne von Notebooks zur Verfügung stehen, mit denen sie das Ergebnis ihrer Recherche festhalten können. Neben den üblichen Browsing-Funktionen enthält eine solche Webseite selbst erzeugte Informationseinheiten. Sie bestehen im wesentlichen aus Bookmarks, Kategorien oder Suchhinweisen. Das hier vorgestellte Konzept ist speziell auf den Zweck der Verschränkung von Perspektiven und der Aushandlung gerichtet; andere Möglichkeiten der

Dokumentation von Web-Recherchen werden z.B. von Torrance [20] und Keller et. al. [12] unterbreitet. Einem Bookmark kann ein selbst gewählter Titel zugeordnet werden. Bei Bedarf (das Sechseck drückt Optionalität aus) kann ein NOTIZ-Symbol hinzugefügt werden, das einen Link zu einer anderen Webseite repräsentiert, auf der der Schüler zum Beispiel ein Exzerpt oder ein Abstract oder Ergänzungen aus anderen Quellen, wie etwa Büchern oder Gesprächen ablegen kann. Neben oder an Stelle von Bookmarks kann es auch Hinweise geben, wie man mit Hilfe einer Suchmaschine geeignete Inhalte finden kann. Es können Kategorien eingeführt werden, um Bookmarks und/oder Suchhinweise zu klassifizieren. Es kann auch Kategorien geben, die zunächst nur zur Vorbereitung einer Inhaltssammlung eingeführt

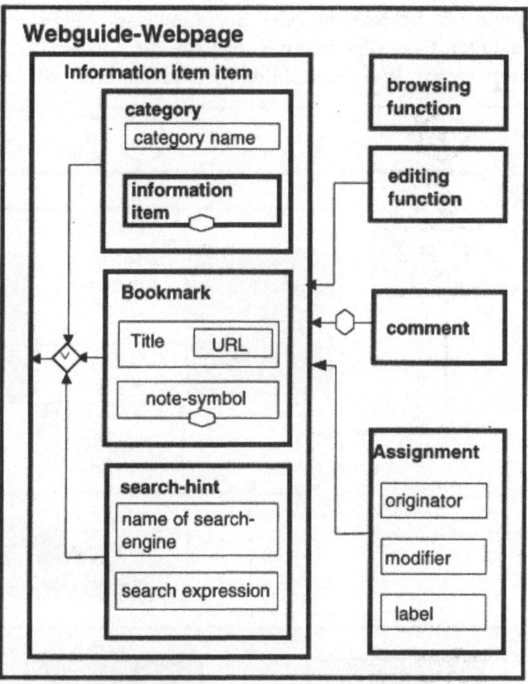

Abb.1   Datenstruktur einer Basis-Webseite in WebGuide
(zur Erläuterung der Symbole s. Abb. 2)

wurden, zu denen es also keine Bookmarks oder Suchhinweise geben muß. Kategorien können hierarchisch gegliedert sein, was durch die rekursive Einbettung der Informationseinheit (im folgenden kurz *Item* genannt) ausgedrückt wird.

Jeder der Informationseinheiten kann bei Bedarf ein Kommentar zugeordnet werden. Es muß zugeordnet werden, wer das Item erzeugt hat, es geändert hat und zu welcher Perspektive es gehört. Zudem ist jedem Item eine Menge von Symbolen für Editierungsfunktionen zugeordnet, von denen aber je nach Eigenart der Informationseinheit nur eine Teilmenge aktiviert werden kann. Eine solche Möglichkeit der Untergliederung in Informationseinheiten ist Voraussetzung für die Möglichkeit der Perspektivenverschränkung durch Aushandlung.

3.2   Das Lösungskonzept im Überblick

Wir stellen uns vor, daß zunächst die zu bearbeitende Aufgabe vom Lehrer (T, siehe Abbildung 2) vorbereitet wird. Dies kann durch die Festlegung der obersten Kategorien geschehen, also etwa entsprechend der vier Indianerstämme. Analog zu ihnen können sich vier Untergruppen in der Klasse bilden. Zudem können auch Unterkategorien und anleitende

Beispiele für Bookmarks oder Suchhinweise gegeben werden. Diese Vorgabe wird in der sogenannten Class-Perspektive abgelegt, die entsprechend der oben beschriebenen Web-Seite strukturiert ist. Wie in Abbildung 2 ersichtlich, können die Schüler (S) danach mit der

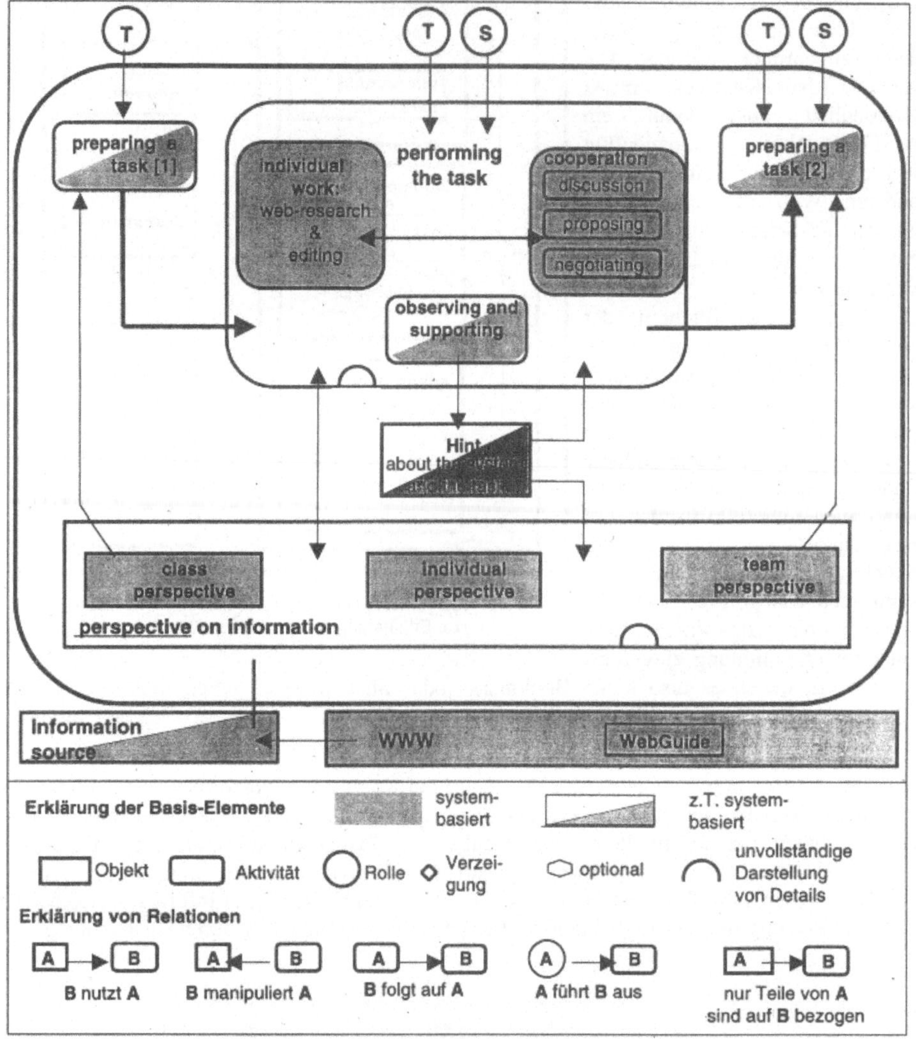

Abb. 2: Grundmodell des WebGuide Konzeptes

Ausführung der Aufgabe beginnen. Sie nutzen dabei die Class-Perspektive, bauen in individueller Arbeit individuelle Perspektiven auf, wobei sie auch auf die Perspektiven anderer Schüler zugreifen können sollen, sobald sie selbst einen Beitrag geleistet haben.

Die kooperative Arbeit besteht im wesentlichen darin, in gemeinsamer Diskussion Informationen zusammenzutragen, Vorschläge für die Team-Perspektive zu unterbreiten, diese Vorschläge zu verhandeln, und die problematischen Fälle wiederum in gemeinsamer Diskussion zu klären. Während der gesamten Aufgabenbearbeitung können der Lehrer, aber auch die Schüler, das Geschehen beobachten und Hinweise geben; unter Umständen ist es sinnvoll, solche Hinweise auch im WWW abzulegen; diese Möglichkeit wird hier nicht näher vertieft. Auf Basis der in der Team-Perpektive zusammengefaßten Ergebnisse kann dann eine weitere Aufgabe vorbereitet werden, wobei die Schüler u.U. mitwirken können.

3.3   Möglichkeiten der individuellen Arbeit

Abbildung 3 zeigt die Aufgabenbearbeitung im Detail. Die Schüler recherchieren in den ihnen zugänglichen Informationsquellen; neben dem WWW können dies natürlich auch Bücher sein, Gespräche oder andere mögliche Quellen, die wir nicht alle antizipieren können, was durch die drei Fragezeichen ausgedrückt ist. Der Schüler kann ebenfalls die Inhalte der Class- und Team-Perspektive sowie der anderen individuellen Perspektiven zur Kenntnis nehmen, die zu diesem Zweck alle in einer Comparison-Perspektive nebeneinander gestellt und überschaubar gemacht werden. Aus der Recherche übernehmen die Schüler diejenigen Daten, die ihnen interessant erscheinen. Zu diesen übernommenen Daten können dann im dritten Schritt eigene Daten, wie etwa Kommentare, eigene Bookmark-Titel etc. hinzugefügt werden. Die drei Schritte können sich beliebig abwechseln und wiederholt werden.

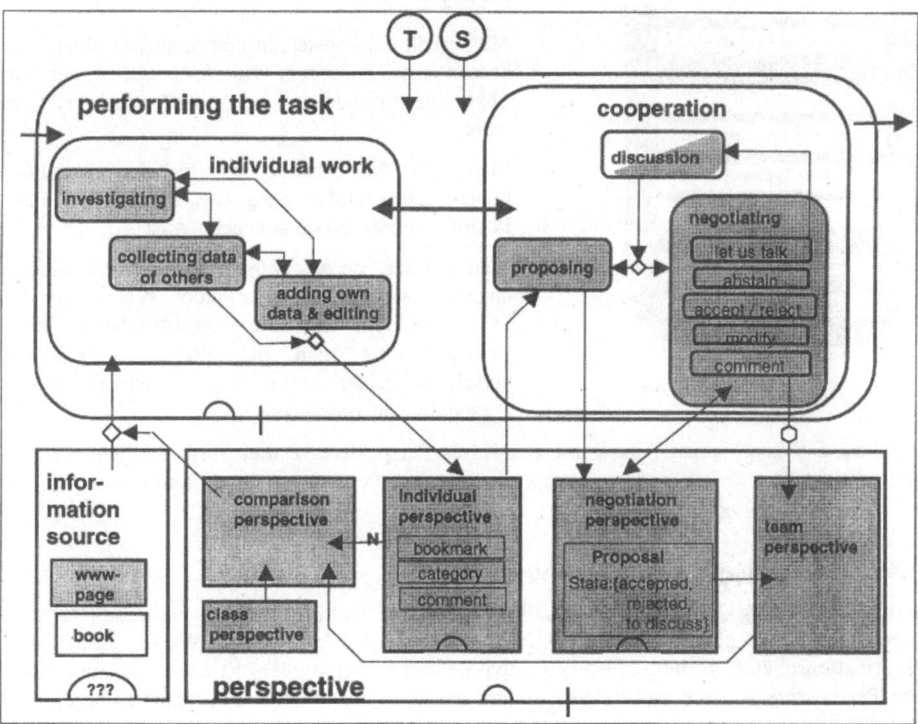

Abbildung 3: Verfeinerte Darstellung der Aufgabenausführung und der Perspektiven

Um die genannten Schritte individueller Arbeit durchführen zu können, ist es sinnvoll, daß man sich zu jedem Item die Menge derjenigen Funktionssymbole anzeigen lassen kann, die in der gegebenen Perspektive auf das Item anwendbar sind. Abb. 4 zeigt die bisher geplante Maximal-Menge von Funktionen.

Der Hide- und Show-Mechanismus ist für jedes Item in jeder Perspektive möglich und bedeutet, daß man sich zum Beispiel nur die Kategorien anzeigen lassen kann, daß man die Kommentare ausblenden kann oder daß man sich alle Details ansieht. Hierdurch soll das Browsen und die Orientierung erleichtert werden. In der Comparison-Perspektive ist kein Editieren möglich; hier hat man nur die Möglichkeit, Items in die eigene Perspektive zu kopieren. Dabei muß es auch möglich sein, Gruppen von Items zu kopieren; man sollte also z.B. nur den Namen einer Kategorie oder auch sämtliche unter einer Kategorie zusammengefaßten Einträge kopieren können.

In der individuellen Perspektive sind verschiedene Editierungsfunktionen möglich, wie Einfügen eines Items hinter die aktuelle Position (1), Verändern von Text (2), Verschieben von Items (3) und Löschen von Items (4). Jede dieser Funktionen bewirkt den Aufruf einer Formularseite, in die weitere Spezifikationen eingetragen werden können:

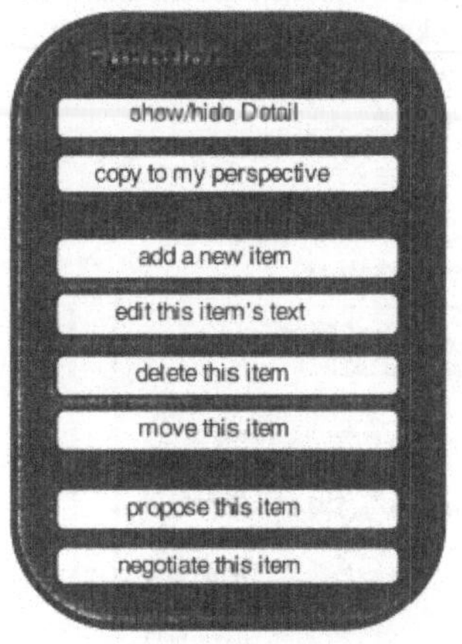

Abbildung 4: Mögliche Funktionen

1. Man kann den Typ des neuen Items festlegen und die Zwischenablage einlesen oder neuen Text eingeben,

2. Man erhält ein Fenster, in dem man den alten Text ändern kann, außerdem wird festgelgt, ob der alte Text durchstrichen angezeigt wird oder gar nicht mehr.

3. Man gibt die Position an, zu der das Item hingeschoben wird und legt fest, ob es an der alten Position durchstrichen angezeigt wird.

4. Man legt fest, ob das gelöschte Item durchstrichen angezeigt werden soll oder nicht. Wenn man die Delete-Funktion auf ein durchstrichenes Item anwendet, hat man die Wahl, es gänzlich verschwinden zu lassen oder es wieder in die Perspektive aufzunehmen.

Abbildung 5 zeigt einen Ausschnitt aus Kay's Perspektive. Daran wird deutlich, daß sie immer sieht, was sie aus anderen Perspektiven in ihre Perspektive übernommen hat und was sie selbst hinzugefügt hat.

3.4 Vorschlagen und Aushandeln

Jeder Schüler kann aus seiner individuellen Perspektive heraus mittels der Funktion Propose Vorschläge für die Team-Perspektive unterbreiten. Auf diese Weise kann eine Informationseinheit in die Team-Perspektive neu eingefügt werden. Er kann auch Items aus der Perspektive anderer vorschlagen, indem er sie zuerst in seine Perspektive übernimmt.

Falls ein bereits bestehender Eintrag der Team-Perspektive geändert werden soll, muß er in eine individuelle Perspektive übernommen werden, dort wird er geändert und dann vorgeschlagen; die Änderung muß nachvollziehbar bleiben, indem mit durchgestrichenem Text gearbeitet wird. Dieser etwas umständliche Weg über die individuelle Perspektive ist sehr sinnvoll, weil auf diese Weise die Änderung vom Vorschlagenden selbst und auch von anderen Schülern genutzt werden kann, bevor das Ergebnis der Aushandlung festliegt. Es muß möglich sein, mehrere zusammenhängende Items in nur **einem** Schritt für die Übernahme vorzuschlagen; z.B.löscht Kay die Kategorie Calendar in der Team-Perspektive und fügt Sun-Calendar neu ein - beide Änderungen sollten zu **einem** Vorschlag zusammengefaßt und in einem Aushandlungsvorgang behandelt werden. Ferner ist es wichtig, daß man Änderungsvorschläge auch dann zusammenfassen kann, wenn die betroffenen Items nicht direkt nebeneinander stehen. Dies bedarf dann mehrerer Schritte. Wenn ein Schüler z.B. etwas an einer Stelle löscht, um an einer anderen Stelle ein besseres, umfassenderes Bookmark einzufügen, dann sollte auch in diesem Fall beides im Zusammenhang vorgeschlagen und auch ausgehandelt werden. Die Schüler sollten allerdings dazu angehalten werden, nicht zu viele Schritte zusammenzufassen.

Wenn die **Proposed**-Funktion aktiviert wird, wird zunächst überprüft, ob der Vorschlag nicht schon exisitiert. Wenn nicht, wird eine WWW-Seite mit Formularfenster geöffnet. Hier kann man festlegen, ob der Vorschlag mit dem vorangegangenen oder mit einem nachfolgenden zusammengefaßt werden soll. Zudem sieht man die Liste aller Schüler, die bei dem unterbreiteten Vorschlag an der Aushandlung beteiligt sein werden. Wer das ist, ist im Rahmen der Aushandlungsphilosophie festzulegen, die möglichst mittels Parameter steuerbar sein sollte, um sie von Fall zu Fall variieren zu können. Beispielsweise kann man dann festlegen, daß Neueinträge immer von allen Teammitgliedern zu verhandeln sind, während bei Änderungen nur diejenigen beteiligt werden, die das zu ändernde Item eingebracht haben bzw. es irgendwann modifiziert haben. Man kann so auch festlegen, daß etwa das Einfügen von Kommentaren gar nicht verhandelt wird.

Actecs [class Pers]

~~Calendar~~ **[Team Pers, proposed by Kay]**
Sun-Calendar [Team Pers, proposed by Kay]

    WEB: <u>Time Units of the Actecs</u> **[Phil's Pers]**
    Comment: Great example **[Kay's Pers]**

    WEB: <u>Religion and Calendar</u> **[Kay's Pers]**

    WEB: <u>A Day in the World of the Actecs</u> **[Team Pers]**
    Comment: Not relevant for calendars **[Kay's Pers]**

Abbildung 5: Informationseinheiten in Kay's Perspektive

Der Vorschlag erscheint dann sowohl in der Team-Perspektive als auch in der individuellen Perspektive aller Verhandlungsteilnehmer mit dem Label „[Team Perspective, proposed by NAME]". Auf das Proposal können zunächst keine Editierungsfunktionen angewendet werden, sondern nur das Aushandeln mittels der **Negotiate**-Funktion. Diese öffnet eine Web-Seite mit einem ersten Fenster, auf dem wahlweise nur der Vorschlag mit oder ohne den

Kontext der Team-Perspektive gezeigt wird. In einem zweiten Fenster kann man die bereits erfolgten Aushandlungsentscheidungen und Kommentare anderer Teammitglieder sehen (s. Tabelle 1).

Zu dem Vorschlag sind nun Aushandlungsreaktionen möglich, wie sie in Abbildung 3 gezeigt werden. Der Aushandelnde kann mit ABSTAIN kundtun, daß er an der Aushandlung nicht teilnehmen möchte. Er kann mit LET US TALK mitteilen, daß er über den Vorschlag in der Gruppe sprechen möchte. Die Folge ist, daß das Label zu „[proposed by NAME, let us talk]" geändert wird. Außerdem kann eine automatisch geführte Agenda von Diskussionspunkten der Gruppenbesprechung automatisch erweitert werden.

Zentral ist es, daß man wählen kann, ob man einen Vorschlag ablehnt oder akzeptiert. Hier ist zu beachten, daß ein unterbreiteter Vorschlag von anderen Gruppenmitgliedern modifiziert werden kann, so daß es für denselben Vorschlag mehrere Alternativen geben kann. In dem obigen, auf Abbildung 5 bezogenen Beispiel, könnte Bea etwa den Vorschlag von Kay modifizieren und als Alternative „Sun-Moon Calendar" vorschlagen Falls ein ACCEPT gewählt wird, werden alle anderen eventuell vorhandenen Alternativen mit einem REJECT versehen und das Aushandlungsfenster geschlossen. Falls REJECT gewählt wurde, wird automatisch die nächste Alternative angezeigt (alle Alternativen sind nacheinander aufgelistet). Beim Angebot mehrerer Alternativen kann man entweder genau eine akzeptieren oder alle zurückweisen.

Nach dem Schließen des Aushandlungsfensters erhält man die Möglichkeit, die eigene Reaktion zu kommentieren. Es wird für alle anderen nachvollziehbar dargestellt, wie man reagiert hat und wie man die Reaktion kommentiert hat. Für den hier dargestellten Fall, daß mehrere Teilnehmer miteinander verhandeln, haben wir entschieden, daß es nicht sinnvoll ist, Reaktionen auf Reaktionen zuzulassen, da sonst der Aushandlungsprozeß sehr schnell unüberschaubar wird. Allerdings gibt es eine Ausnahme: alle Reaktionen, auch die von anderen, können kommentiert werden. Auch wenn man sein Votum schon abgegeben hat, kann man immer wieder das Aushandlungsfenster öffnen, um Reaktionen anderer zu kommentieren oder das eigene Votum zu ändern. Diese Funktion wird pro dargestellter Reaktion im Kommentarfenster angeboten - siehe als Beispiel Tabelle 1. Konsequenterweise kann man auch Kommentare kommentieren.

Name	Reaktion	Statistik
Que:	acceptance of Kay's proposal	1 of 3
	Comment: I think this makes sense	
Bea	rejection of Kay's proposal and modification of Kay's proposal	1 of 1
Phil	rejection of Bea's modification	1 of 1
Jay	rejection of Bea's modification	1 of 2
	Comment: I prefer the original	
Bea	comment on Jay's comment	1 of 1
	Comment: why do you prefer it?	

Tabelle 1: Beispiel für die Darstellung von Reaktionen und Kommentaren

Die Möglichkeit der **Modifikation** wird indirekt angeboten. Sobald alle verfügbaren Alternativen eines Aushandlungsvorschlages abgelehnt wurden, kann der Ablehnende in seiner individuellen Perspektive den Vorschlag editieren und ihn in geänderter Form vorschlagen. Diese Alternative wird dann automatisch dem Aushandlungsvorgang hinzugefügt. Ebenso wird das Label geändert „[proposed by NAME1, altered by NAME2]. Die Aushandelnden sehen dann die Alternativvorschläge und können entscheiden, ob sie noch einmal reagieren wollen, um ihr Votum zu ändern. Vor diesem Hintergrund ist es auch sinnvoll, einen Ereignisdienst einzuführen, der auf solche Änderungen hinweist. Die Design-Entscheidung, die Modifikation nur auf diese Weise zu ermöglichen, führt zu einer Vereinfachung, da der Schüler immer nur unter den gleichen Bedingungen editiert, die er im Kontext seiner individuellen Perspektive kennt. Es ist nur möglich, den Ursprungsvorschlag zu ändern, nicht aber proposed Items, die bereits das Label *altered* haben, da sonst die Übersichtlichkeit verloren ginge. Ein einmal gemachter Vorschlag kann zwar vom Urheber per Votum abgelehnt werden, wenn ihm eine Alternative besser gefällt, der Vorschlag kann aber nicht ganz zurückgenommen werden, da er ggf. der Präferenz anderer Teilnehmer entspricht.

Das Aushandlungsgeschehen kann pro Vorschlag nur über einen im Rahmen der Aushandlungsphilosophie festgelegten Zeitraum erfolgen. Nach Ablauf der Zeit wird festgestellt, ob der Vorschlag bzw. eine der Alternativen akzeptiert oder abgelehnt ist. Für diese automatische Feststellung wird vorab spezifiziert, welcher Mehrheitsmodus (Zwei-Drittel, absolute oder einfache Mehrheit etc.) anzuwenden ist und wie Ablehnung gegen Akzeptanz aufzurechnen ist. Falls kein eindeutiges Votum zustande kommt, wird der Vorschlag mit *proposed for talk* gekennzeichnet und auf die Besprechungsagenda genommen. Die Schüler müssen dann versuchen, sich in direktem Gespräch über den Fall zu einigen. Während solcher Besprechungen sollte es möglich sein, Änderungen direkt in der Team-Perspektive vorzunehmen, indem man sich eines Super-Passwortes bedient, zu dem alle einen Teil-String beitragen. So kann das Ergebnis der Gruppendiskussion unter Umgehung der systembasierten Aushandlung eingegeben werden.

4 Zusammenfassung

Es wurde deutlich, daß das hier vorgeschlagene Konzept nur dann umsetzbar ist, wenn es gelingt, den zu bearbeitenden Informationsraum kleingliedrig zu zerlegen. Der Ansatz läßt sich zum Beispiel kaum auf ein Shared Editing längerer Text anwenden. Ferner ergeben sich einige software-ergonomische Probleme: Man kann mit WWW-Seiten selbst unter Nutzung von JAVA das Prinzip der direkten Manipulation nicht immer konsequent verwirklichen, da es immer wieder darum geht, daß nach jeder Änderung neue Seiteninhalte erzeugt werden, die auch anderen zugänglich sind.

Es gibt einen Protoypen, der einige Perspektiven des Systems zeigt, ohne daß die dazugehörige Funktionalität verfügbar ist[1]. Man erkennt aber das Konzept der WWW-Seiten-Gestaltung und die zentralen Entscheidungen, die hinsichtlich der Informationsdarstellung getroffen wurden. Das System wird weiter vervollständigt und wird in 1998 in einer

---

[1] Siehe http://www.cs.colorado.edu/~gerry/WebGuide/webguide.htm

110

konkreten Schulumgebung evaluiert werden. Dabei wird es darum gehen, zum einen das Nutzungsverhalten zu evaluieren. Zum Beispiel: Wie wird das Zusammenfassen von Vorschlägen zu einem Aushandlungsvorgang genutzt; wie stark wird das Kommentieren genutzt, wie gehen Schüler mit der Möglichkeit um, Vorschläge zu modifizieren? Zum anderen sollen Hinweise für die Aushandlungsphilosophie gewonnen werden: Wie verfährt man mit Vorschlägen, die nur von einem oder zweien zur Besprechung vorgeschlagen werden? Welche Zeitlimits, Abstimmungsmodi, Teilnehmer der Aushandlung etc. sind sinnvoll?

Es wurde deutlich, daß das ursprüngliche Aushandlungskonzept stark vereinfacht werden mußte, um für Schüler handhabbar zu sein. Betrachtet man die hier vorliegende textliche Darstellung, so wirkt das Konzept immer noch recht komplex. Die Autoren gehen davon aus, daß sich die Möglichkeiten einfacher darstellen, wenn man sie am System direkt erkunden kann. In der Evaluationsphase können sich sicherlich Hinweise für weitere Vereinfachungsmöglich-keiten ergeben, und es wird sich zeigen, was unter Umständen in einer Systemerklärung bereitgestellt werden muß. Wenn es gelingt, das Konzept für Schüler ab 12 Jahren ausreichend handhabbar und überschaubar zu gestalten, dann lassen sich an diesem Beispiel wichtige Hinweise gewinnen, wie man das Prinzip verschränkter Perspektiven durch Aushandlung auch für andere Anwendungsfelder benutzbar gestalten kann.

LIteratur

[1]   Ackerman, M. S.: Augmenting the Organizational Memory: A Field Study of Answer Garden. In: The Conference on Computer Supported Collaborative Work(CSCW'94), New York, 1994: ACM. S. 243-252.

[2]   Bayer, Elke: Flexibilität in E-Mail Systemen - Ein Modell für Aushandelbarkeit. Diplomarbeit an der Universität Bonn, 1995.

[3]   Bieber, M., Vitali, F., Ashman, H., Balasubramanian, V., Oinas-Kukkonen, H.: Fourth generation hypermedia: Some missing links for the World Wide Web. International Journal of Human-Computer Studies. Special issue on HCI & the Web, 1997.

[4]   Boland, R. J., Maheshwari, A. K., Te'eni, D., & Tenkasi, R. V.: Sharing Perspectives in Distributed Decision Making. In: The Conference on Computer Supported Collaborative Work (CSCW'92). New York, 1992: ACM. S. 306-313.

[5]   Ephrati, Eithan; Zlotkin, Gilad; Rosenschein, Jeffrey: Meet Your Destiny: A Non-manipulable Meeting Scheduler. In: Proceedings of the CSCW '94. New York, 1994: ACM. pp. 359-371.

[6]   Fitzgerald F., Rashid R.: The Integration of Virtual Memory Management and Interprocess Communication in Accent. ACM Transactions on ComputerSystems, 1986, 4, (2), 147.

[7]   Floyd, Christiane: Software Development and Reality Construction. In: Floyd, Christiane; Züllinghoven, Heinz; Budde, Reinhard; Keil-Slawik, Reinhard: Software Development and Reality Construction. Berlin, Heidelberg, New York, 1992: Springer-Verlag. S. 86-100.

[8]   Geibel, Richard: Computergestützte Gruppenarbeit. Die Förderung von Gruppenentscheidungen durch Group Decision Support System. Stuttgart, 1993: M&P Verlag.

[9]   Habermas, Jürgen: Theorie des kommunikativen Handelns. Band 1. Handlungsrationalität und gesellschaftliche Rationalisierung. Frankfurt, 1981: Suhrkamp.

[10]   Herrmann, Thomas: Grundsätze ergonomischer Gestaltung von Groupware. In: Hartmann, Anja; Herrmann, Thomas; Rohde, Markus; Wulf, Volker (Hrsg.): Menschengerechte Groupware - Software-ergonomische Gestaltung und partizipative Umsetzung. Stuttgart, 1994: Teubner Verlag. S. 65-107.

[11]   Herrmann, Thomas: Workflow Management Systems: Ensuring organizational Flexibility by Possibilities of Adaption and Negotiation. In: Comstock, Nora et al. (Hrsg.): COOCS´95. Conference on Organizational

Computing Systems. August 13-16, 1995. Milpitas, California. New York: acm-press. New York, 1995: ACM. S. 83 - 95.

[12] Keller, R., Wolfe, S.; Chen, J.; Rabinowitz, J.; Mathe, N.: A bookmarking service for organizing and sharing URLs. In: Computer Networks ans ISDN Systems 29, 1997. S. 1103-1114.

[13] Koschmann, Timothy (ed.): CSCL: Theory and Practice. New Jersey, 1995: Lawrence Erlbaum Associates.

[14] Lindstaedt, Stefanie; Schneider, Kurt: Bridging the Gap between Face-to-Face Communication and Long-Term Collaboration. In: Hayne, S.; Pronz, W. (eds.): Proceedings of the International ACM SigGroup Conference on Supporting Group Work. The Integration Challenge. New York, 1997: ACM. S. 331-340.

[15] Martial, Frank von: A Conversation Model for Resolving Conflicts among Distributed Office Activities. In: SIGOIS 2,3. 1996. S. 99 - 108.

[16] Resnick, Paul; Iacovou, Neophytos; Suchak, Mitesh; Bergstrom, Peter: GroupLens: An Open Architecture for Collaborative Filtering of Netnews. In: Proceedings of the CSCW '94: Transcending Boundaries. NY, 1996: ACM Press. S. 175 -186.

[17] Sen, Sandip; Haynes, Thomas; Arora, Neeraj: Satisfying user preferences while negotiating meetings. In: Int. J. Human-Computer Studies. 1997. S. 407-427.

[18] Stahl, Gerry: Interpretation in Design: The Problem of Tacit and Explicit Understanding in Computer Support of Cooperative Design. University of Colorado at Boulder, 1993.

[19] Stefik, M.; Foster, D.; Bobrow, D.G.; Kahn, K.; Lanning, S.; Suchman, L.: Beyond the Chalkboard: Computer Support for Collaboration and Problem Solving in Meetings. In: Greif, Irene (ed.): Computer-Supported Cooperative Work. 1988, S. 335-366.

[20] Torrance, M.: Active notebook: A personla and group productivity tool for managing information. In: Proc. AAAI Fall Symposion on Artificial Intelligence Applications in Knowledge Navigation and Retrieval. Cambridge Massachusetts. 1995, S. 131-135.
(http://www.ai.mit.edu/people/torrance/papers/aaai-fall-95.ps)

[21] Wulf, Volker: Konfliktmanagement bei Groupware. Braunschweig, Wiesbaden, 1997: Vieweg

Adressen der Autoren

Prof. Dr.-Ing. Thomas Herrmann
Universität Dortmund
Fachgebiet Informatik & Gesellschaft
FB Informatik Lehrstuhl VI
D - 44221 Dortmund
E-mail: herrmann@iug.cs.uni-dortmund.de

Gerry Stahl, Ph.D.
University of Colorado, Boulder
Department of Computer Science
Center of LifeLongLearning and Design
Boulder, CO 80309-0430
E-mail: Gerry.Stahl@colorado.edu

Gruppenwahrnehmung und Kommunikation bei Web basierten Kooperationswerkzeugen

Thomas Koch, Wolfgang Appelt

Forschungsbereich Kooperationssysteme, GMD Forschungszentrum Informationstechnik

Zusammenfassung

In diesem Beitrag wird beschrieben, wie mit Hilfe von Java Applets bei WWW basierten Systemen Möglichkeiten für synchrone Kommunikation sowie Awareness-Informationen über die Präsenz und Aktivitäten von Web Benutzern bereitgestellt werden können. Insbesondere wird aufgezeigt, wie diese Funktionalität in das BSCW Shared Workspace System, einem in der GMD entwickeltem Web basierten Groupware System, integriert worden ist.

1 Einleitung

Das World Wide Web ist ein leicht zu benutzendes, aber mächtiges globales Informationssystem, das auf zwei einfachen Standards basiert, dem HyperText Transfer Protocol (HTTP) und der HyperText Markup Language (HTML). Es besteht aus vielen unabhängigen Servern, die von den Endbenutzern über Netze mit Web Browsern (Clients) kontaktiert werden und auf Aufforderung entsprechende Antworten an die Browser senden. Zwar ist die Bereitstellung und Verteilung von Informationen die am häufigsten übliche Nutzung des Web, doch sind inzwischen auch schon eine Reihe von CSCW (Computer Supported Cooperative Work) Anwendungen entwickelt worden, die mit Konzepten wie gemeinsamen virtuellen Arbeitsräumen mehr interaktive und kommunikative Nutzungsweisen ermöglichen. Insbesondere für lokal verteilte Gruppen - möglicherweise mit Mitgliedern aus unterschiedlichen Organisationen - bietet es sich an, Kooperationsdienste auf WWW Basis zur Verfügung zu stellen. Zu den wesentlichen Vorteilen des WWW und den Gründen für den Einsatz von CSCW Systemen auf Basis des WWW zählen die kritische Masse an Benutzern, die Integration vorhandener Anwendungen und die breite plattformübergreifende Softwareunterstützung [4].

Die Entwicklung von CSCW Anwendungen auf Web Basis bringt allerdings auch Probleme mit sich. Das verwendete Client/Server Konzept sowie HTTP als Protokoll sind im Prinzip nur für asynchrone Kommunikation geeignet. So überträgt etwa der Server nur dann Informationen zum Client, wenn dieser explizit danach fragt. Wenn auf dem Server Änderungen erfolgen (zum Beispiel die Modifikation einer HTML Seite), kann dieser nicht von sich aus Clients davon informieren. Außerdem können Clients wie etwa Web Browser auf HTTP Basis nicht direkt miteinander kommunizieren, sondern nur indirekt und in der Funktionalität sehr begrenzt über einen Server.

Zur Lösung dieser Probleme könnte man das HTTP Protokoll so ändern und erweitern, daß es genügend Unterstützung für synchrone Kommunikation bietet. Da dies Konsequenzen für alle derzeit vorhandenen HTTP Anwendungen haben dürfte, ist es zweifelhaft, ob sich dieser Weg durchsetzen ließe und er wäre sicher nur in einem längerem Zeitraum durchführbar. Ein zweiter Weg besteht darin, daß man die Web Clients so modifiziert, daß sie synchrone Kommunikation unterstützen. Dieser Ansatz wird auch häufig gewählt (wie zum Beispiel bei NetMeeting von Microsoft oder Conference von Netscape), hat aber - neben der Tatsache, daß die Benutzer zusätzliche Software installieren müssen - den Nachteil, daß plattformübergreifende Kommunikation in aller Regel nicht möglich ist und damit eines der attraktivsten Eigenschaften des World Wide Web, nämlich seine Plattformunabhängigkeit, verloren geht.

Ein dritter Ansatz, der die Nachteile der beiden ersten vermeidet, besteht darin, daß man Java-Applets zur Unterstützung synchroner Kommunikation verwendet. Da heute praktisch alle Browser Java unterstützen, ist diese Lösung plattformunabhängig und erfordert nur Änderungen auf Serverseite, aber keine Änderung des HTTP Protokolls. Nachfolgend wird beschrieben, wie wir mit diesem Ansatz Unterstützung zur synchronen Kommunikation für das BSCW System - einem Groupware System, das im nächsten Absatz etwas detaillierter beschrieben wird - integriert haben.

2 Das BSCW System

Der Ansatz des BSCW Systems (BSCW = Basic Support for Cooperative Work) ist es, Kommunikation und Kooperation von Benutzern auf der Grundlage von sogenannten gemeinsamen Arbeitsbereichen (*Shared Workspaces*) zu unterstützen. Die Mitglieder einer Arbeitsgruppe richten solche Arbeitsbereiche auf einem BSCW Server ein und verwenden sie zur Organisation und Koordinierung ihrer Aufgaben. Ein solcher Arbeitsbereich kann unterschiedliche Arten von (elektronischen) Objekten wie zum Beispiel Dokumente, Tabellen, Grafiken oder Verweise auf WWW Seiten enthalten. Die Mitglieder der Arbeitsgruppe können Objekte von ihrem lokalen Rechner auf den Arbeitsbereich übertragen (*Upload*) oder Objekte vom Arbeitsbereich auf ihre lokalen Rechner transferieren (*Download*), etwa um ein Dokument zu lesen oder zu editieren. Anders formuliert: Ein BSCW Shared Workspace ist zunächst einmal ein zentraler Informationsspeicher, ähnlich einem Ftp-Archiv.

Darüber hinaus stellt das BSCW System jedoch eine Reihe weiterer Funktionen zur Verfügung, die die Kooperation der Mitglieder einer Arbeitsgruppe unterstützen. Zum Beispiel registriert das System alle sogenannten Ereignisse (*Events*) in einem Arbeitsbereich. Ein Ereignis wird dabei durch jeden Zugriff auf einen Arbeitsbereich ausgelöst, zum Beispiel wenn ein neues Objekt abgelegt wird, eine neue Version eines existierenden Objekt erzeugt wird, ein *Download* eines Dokuments erfolgt oder ein Objekt umbenannt wird. Wenn ein Benutzer einen BSCW Shared Workspace "betritt", wird er darüber informiert, welche Ereignisse sich in der letzten Zeit ereignet haben, das heißt, jedes Mitglied einer Arbeitsgruppe erhält auf diese Weise Informationen über die Aktivitäten der anderen Mitglieder im Arbeitsbereich. Der Begriff "in der letzten Zeit" bedeutet hier, daß diejenigen Ereignisse angezeigt werden, die sich ereignet haben, nachdem der Benutzer zum letzten Mal eine sogenannte "Bestätigen"-Aktion ausgeführt hat. Mit dieser Aktion teilt der Benutzer dem BSCW System mit, daß er ein oder mehrere Ereignisse zur Kenntnis genommen hat und diese ihm ab sofort nicht mehr als "neu" angezeigt werden sollen.

Im Kontext von Groupware-Anwendungen spricht man dabei auch von gegenseitigem Bewußtsein (*mutual awareness*). Der Begriff wird von Dourish und Bellotti [5] wie folgt erklärt:

„... awareness is an understanding of the activities of others, which provides a context of your own activity. "

Erst das Bewußtsein über die Aktivitäten anderer Mitarbeiter erlaubt dem Gruppenmitglied, die eigene Arbeit sinnvoll zu strukturieren und es entsteht ein gemeinsamer Kontext des Arbeitsprozesses.

Darüber hinaus bietet das BSCW System eine Reihe weiterer Funktionen:

- *Authentifizierung*: Benutzer müssen sich mit einem Namen und Paßwort identifizieren, bevor sie Zugang zu ihren Arbeitsbereichen erhalten.

- *Versionsverwaltung*: Dokumente in Arbeitsbereichen können unter eine mit RCS vergleichbare Versionsverwaltung gestellt werden, was insbesondere für gemeinsame Dokumenterstellung hilfreich ist.

- *Diskussionsforen*: Benutzer können ähnlich wie in Internet Newsgroups Diskussionen führen.

- *Zugriffsrechte*: BSCW enthält ein recht ausgefeiltes Modell bezüglich der Zugriffsrechte auf Objekte in Arbeitsbereichen, mit dem man zum Beispiel spezifizieren kann, daß bestimmte Benutzer nur Leserechte oder überhaupt keine Zugriffsrechte auf bestimmte Objekte besitzen.

- *Suchfunktionen*: Die Benutzer können nach Objekten in BSCW Arbeitsbereichen suchen, wobei sich die Anfrage auf Namen, Inhalt oder sonstigen Objekteigenschaften (wie etwa Name des Autors oder Erstellungszeitraum) eines Dokuments beziehen kann. Es lassen sich auch Anfragen an externe Suchmaschinen im WWW absetzen, bei denen die von den Suchmaschinen zurückgelieferten Treffer unmittelbar in BSCW Objekte transformiert werden.

- *Dokumentkonversion*: Diese Funktionen gestatten es dem Benutzer, vor dem *Download* eines Dokuments dieses in ein anderes Format zu konvertieren, zum Beispiel ein MS Word Dokument in ein HTML Dokument, weil etwa der Benutzer MS Word Dokumente nicht bearbeiten kann.

- *Konfigurierbarkeit*: Der Benutzer kann in bestimmtem Umfang die Benutzer-schnittstelle des Systems nach seinen Bedürfnissen anpassen, zum Beispiel spezifizieren, ob er Javascript oder ActiveX Erweiterungen verwenden will, oder aus einer Reihe von verschiedenen Sprachen diejenige auswählen, in der ihm die vom System erzeugten HTML Seiten präsentiert werden sollen.

Die Implementation des Systems folgt dem üblichen Client/Server-Modell: Auf dem *BSCW Server* sind die BSCW Arbeitsbereiche abgelegt. Als *BSCW Client* fungiert ein Web Browser, über den die Benutzer mit dem BSCW Server kommunizieren. In Abbildung 1 ist ein Beispiel zu sehen, wie sich dem Benutzer ein Ausschnitt aus einem BSCW Arbeitsbereich präsentiert. (Eine detaillierte Beschreibung des BSCW Systems ist in [2] zu finden.)

Der BSCW Server ist eine Erweiterung üblicher WWW Server, zum Beispiel Apache oder der Windows NT Web Server. Die Erweiterung erfolgt dabei durch das Common Gateway Interface (CGI) des verwendeten Web Servers. Die Software selbst ist in Python[1] geschrieben,

---

[1] Python Home Page - http://www.python.org/

einer interpretierten und objektorientierten Sprache, für die Interpreter kostenlos für zahlreiche Plattformen verfügbar sind.

Die GMD betreibt einen allgemein zugänglichen BSCW Server[2], wo Interessierte kostenlos Arbeitsbereiche einrichten und zur Kooperation mit Kollegen benutzen können. Von dort ist auch die Software für nicht-kommerzielle Zwecke kostenlos erhältlich. Von diesen Angeboten wird von zahlreichen Personen Gebrauch gemacht; wir schätzen, daß die derzeitige Benutzerzahl des System weit über 10.000 beträgt.

3   Synchrone Awareness und Kommunikation in BSCW

Wie im vorherigen Abschnitt schon erklärt wurde, können sich die Benutzer des BSCW Systems mittels einer Ereignishistorie über Aktivitäten in einem Arbeitsbereich informieren, die seit der letzten Bestätigung stattgefunden haben. Der Benutzer muß dazu allerdings explizit die Informationen anfordern; treten während seiner Tätigkeit im Arbeitsbereich Aktivitäten anderer Benutzer auf, so wird er dieser nicht automatisch gewahr. In der Regel wird ein Benutzer es sogar nicht einmal bemerken, wenn gleichzeitig andere Benutzer einen Arbeitsbereich inspizieren, da das Browsen im Web üblicherweise in völliger Isolation vom Browsen anderer Benutzer erfolgt. Dadurch fehlt eine der Grundvoraussetzungen für den Einsatz synchroner Kommunikationsanwendungen, nämlich die Information über die zeitgleiche Präsenz anderer Benutzer und damit deren potentielle „Erreichbarkeit".

Daher wurde das BSCW System um eine weitere, im wesentlichen externe Komponente erweitert, die eine Gruppenwahrnehmung (*team awareness*) der Benutzer bei gleichzeitiger Arbeit mit dem System unterstützt. (Die Komponente ist weitgehend extern zur eigentlichen BSCW Software, da diese HTTP als Kommunikationsprotokoll verwendet und mit HTTP die angestrebte Funktionalität nicht erreicht werden kann.) Deren Funktionalität umfaßt im Wesentlichen die folgenden drei Bereiche:

1. Information über Präsenz anderer Benutzer,

2. direkte Benachrichtigung über Aktivitäten in einem BSCW Arbeitsbereich und

3. synchrone Kommunikation zwischen Benutzern.

Die erweiterte Funktionalität wird dem Benutzer über ein zusätzliches Anwendungsfenster bereitgestellt, das der Benutzer nach Anmeldung beim BSCW System über einen entsprechenden Menüpunkt öffnet. Im Fenster befindet sich mit dem Monitor (s. Abbildung 1) ein Java Applet, das eine Übersicht über alle derzeit anwesenden Mitarbeiter liefert. Angezeigt werden dabei alle Benutzer mit denen der jeweilige Benutzer einen gemeinsamen Arbeitsbereich teilt und die ebenfalls den Monitor gestartet haben.

Neben der Übersicht der anwesenden Mitarbeiter bietet der Monitor eine Benachrichtigung über aktuelle Vorgänge in den Arbeitsbereichen des BSCW Systems. Wird z.B. durch einen Benutzer im BSCW ein Dokument erzeugt oder verändert, so wird dies unmittelbar im Monitor angezeigt. Der Benutzer kann dann über den Monitor direkt zum entsprechenden Verzeichnis im Arbeitsbereich wechseln und das betroffene Dokument öffnen oder genauere Informationen über das aufgetretene Ereignis abfragen (s. Abbildung 2).

---

[2] BSCW Home Page - http://bscw.gmd.de/

Über die Liste der aktiven Benutzer erhält man weitere Informationen über einen Benutzer oder tritt mit diesem in Kontakt. Dazu kann man entweder eine email versenden oder dem Benutzer eine kurze Nachricht schicken, die sofort auf dessen Bildschirm erscheint.

Außerdem besteht die Möglichkeit, über gleichzeitige textuelle Kommunikation (chat) eine Diskussion mit mehreren Benutzern durchzuführen.

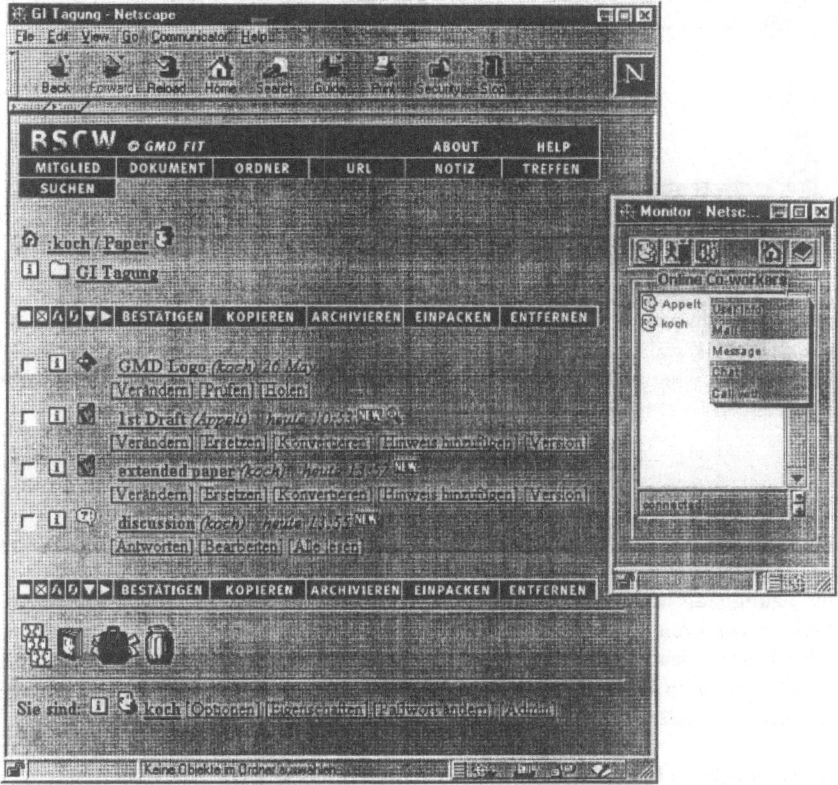

Abbildung 1: BSCW Arbeitsbereich und Monitor

Will man spezielle Kommunikationskanäle verwenden, so kann man über den Monitor eine Audio/Video-Verbindung zu einem anderen Benutzer aufbauen. Allerdings sind diese Anwendungen plattformspezifisch und müssen beim jeweiligen Benutzer lokal installiert sein. Daher bietet das BSCW-System jedem Benutzer die Möglichkeit, anzugeben, welche Kommunikationsanwendungen ihm zur Verfügung stehen. Der Monitor benutzt diese Liste, um einem Anrufer die möglichen Verbindungsarten zum Anzurufenden anzuzeigen. Zum Aufbau einer Verbindung startet der Monitor die entsprechende Anwendung und versorgt sie mit den benötigten Parametern (zum Beispiel mit der IP-Adresse des Anzurufenden).

Der Monitor wurde so konzipiert, daß er dem Benutzer einerseits die gewünschten Zusatzinformationen über erreichbare Mitarbeiter und aktuelle Vorgänge bietet und andererseits keine große Ablenkung von der eigentlichen Tätigkeit (Interaktion mit BSCW,

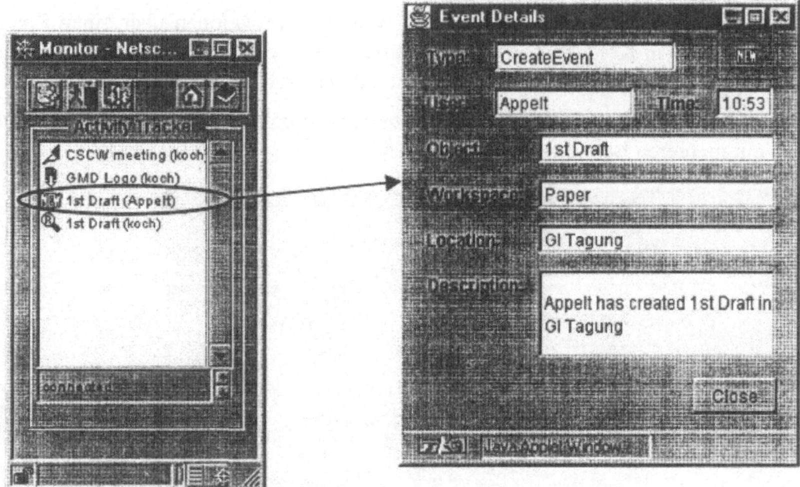

Dokumentbearbeitung etc.) darstellt. Durch die beschränkte Ausdehnung und die gewählte Benutzerschnittstelle kann der Benutzer den Monitor so auf seinem Desktop plazieren, daß keine Beeinträchtigung der Arbeit stattfindet und gleichzeitig eine periphere Wahrnehmung wichtiger Ereignisse möglich ist.

Abbildung 2: Ereignisübersicht und Detailansicht

Darüber hinaus kann der Monitor so konfiguriert werden, daß zum Beispiel keine Benachrichtigung über das Lesen von Dokumenten auftritt, denn diese Ereignisse treten relativ häufig auf und sind oft nicht von Interesse. Außerdem kann die audiovisuelle Darstellung der Ereignisse (in weiterem Anwendungsfenster, mit/ohne Audiounterstützung etc.) durch den Benutzer verändert werden. Die Erreichbarkeit für die Kommunikation mit anderen Mitarbeitern kann ebenfalls durch den Benutzer ausgeschaltet werden, z.B. während eine Arbeitspause oder einer Arbeitsphase, in der eine Störung unerwünscht ist.

4 MetaWeb - Unterstützung synchroner Gruppenarbeit

Bei der Implementation der im vorherigen Abschnitt beschriebenen Zusatzfunktionalität für das BSCW System haben wir uns nicht nur auf die konkrete Anwendung für BSCW beschränkt, sondern eine generische Entwicklungsumgebung zur Konstruktion synchroner Kooperationswerkzeuge - das sogenannte MetaWeb System - entwickelt. Der oben beschriebene Monitor im BSCW System läßt sich somit als ein Anwendungsbeispiel auffassen, das mit dem MetaWeb System realisiert werden kann.

Zentrale Idee des MetaWeb Modells ist die Bereitstellung von Meta Informationen über aktuelle Vorgänge in Web basierten Anwendungen. Dazu basiert das Modell auf folgenden abstrakten Konzepten: Ort, Konferenz, Benutzer und Ereignis. Ein Ort beschreibt den Rahmen einer Aktivität und kann etwa eine WWW Seite aber auch eine anwendungsspezifische Domäne beschreiben. An einem Ort können beliebig viele Konferenzen stattfinden. Eine Konferenz stellt zum einen eine (dynamische) Gruppe dar und ist zum anderen Ausgangspunkt für die synchrone Kooperation in der Gruppe, was Ellis [6] wie folgt als *cooperative session* definiert:

"a period of synchronous interaction supported by a groupware system. Examples include formal meetings and informal work group discussions."

Die Benutzer bilden als Konferenzteilnehmer eine Gruppe und können an beliebig vielen Konferenzen teilnehmen. Ereignisse dienen zur Beschreibung von Zustandsänderungen in verteilten Systemen und erlauben die Verwaltung eines gemeinsamen Zustandes sowie den Nachrichtenaustausch zwischen verteilten Anwendungen.

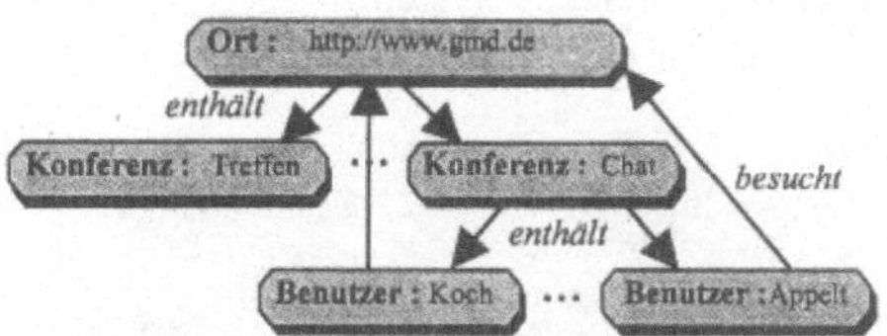

Abbildung 3: Beispiel der Konzepte Ort, Konferenz und Benutzer

Während die Begriffe Konferenz (oder *session*) und Benutzer in Groupware Anwendungen weit verbreitet sind, wurde der Ortsbegriff hier neu eingeführt, um eine flexible Kopplung von Kooperationsprozessen an bestimmte Anwendungsbereiche zu ermöglichen. Wie in Abbildung 3 gezeigt, ist es so etwa möglich, kooperative Prozesse mit WWW Seiten zu verknüpfen.

Das MetaWeb System bietet dem Anwendungsentwickler eine Grundfunktionalität, die die Entwicklung synchroner Kooperationswerkzeuge erleichtern soll. Dazu zählen im Wesentlichen der Zugriff auf folgende Dienste:

- gemeinsamer Informationsraum (*shared objects service*)

- Benachrichtigungsdienst (*notification service*)

- Konferenzverwaltung (*session management*)

Den grundlegenden Konzepten des MetaWeb Modells entsprechen Objekttypen im MetaWeb System. So wird in einer verteilten MetaWeb Anwendung jeder Benutzeragent durch ein entsprechendes **user** Objekt repräsentiert. Anwendungen können Orte und Konferenzen durch Objekte vom Typ **location** und **session** erzeugen, verändern sowie zueinander in Relation setzen. Das MetaWeb System übernimmt dabei die Verwaltung der Objekte im gemeinsamen Informationsraum sowie die Benachrichtigung interessierter Anwendungen bei Zustandsänderungen. Wird ein Objekt verändert, so informiert der Benachrichtigungsdienst die betroffenen Anwendungen über ein entsprechendes Ereignis (Objekttyp **event**).

Das Interesse an Zustandsänderungen wird durch das System ermittelt. Anwendungen können dazu Interesse an bestimmten Objekten beim Benachrichtigungsdienst registrieren. Außerdem können Anwendungen Ereignisse direkt an einen Benutzer (und damit an die ihn vertretende Anwendung) oder an eine Konferenz (und damit an eine Gruppe von Benutzern) versenden.

MetaWeb basiert wie das Web auf einer Client/Server Architektur. Ein zentraler MetaWeb Server verwaltet die Meta Informationen und ermöglicht die Kommunikation zwischen MetaWeb Clients. Diese bieten dem Anwendungsentwickler über eine Programmierschnittstelle (API, *application programming interface*) einen Zugang zur Funktionalität des MetaWeb Systems (s. Abbildung 4).

Der MetaWeb Client kommuniziert über ein spezielles Protokoll mit dem MetaWeb Server. Dieses bleibt der Anwendung durch die Programmierschnittstelle verborgen, die eine höhere Abstraktionsebene als das verwendete Protokoll darstellt.

Durch die Verwendung mobiler Anwendungen wird eine dynamische, plattformunabhängige Client Erweiterung erreicht. Sowohl MetaWeb Client als auch die darauf basierende Anwendung werden durch den Web Client geladen und ausgeführt. Dies ermöglicht eine transparente Erweiterung des WWW durch das MetaWeb System.

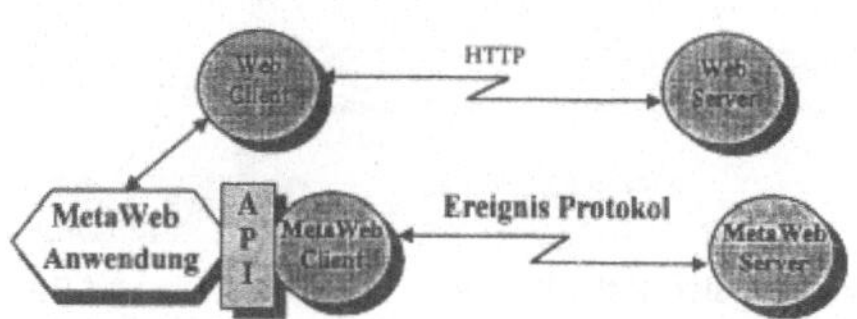

Abbildung 4: Client/Server-Architektur von MetaWeb

Das MetaWeb System wurde in Form einer Java Klassenbibliothek implementiert, da die Programmiersprache Java [1] die geforderte dynamische Client Erweiterung durch Java Applets erlaubt und inzwischen von den am weitesten verbreiteten Browsern (Netscape Communicator und MS Internet Explorer) unterstützt wird. Dies ermöglicht insbesondere – wie auch im BSCW System – eine plattformübergreifende Kommunikation der Clients. Weitere Gründe, die für den Einsatz von Java sprechen, ist die Unterstützung objektorientierter Paradigma, das vorhandene Sicherheitskonzept und die breite Entwicklungsunterstützung durch die vielfältigen Systembibliotheken.

Die MetaWeb Klassenbibliothek realisiert den MetaWeb Server und die Programmierschnittstelle. Der Server arbeitet multi-threaded, verwaltet Verbindungen zu angemeldeten Anwendungen, sorgt für die Verteilung von Ereignissen und verwaltet die Objekte des gemeinsamen Informationsraumes. MetaWeb Client und Server kommunizieren über den Austausch von Ereignissen, wobei zwischen systemspezifischen und anwendungsspezifischen Ereignissen unterschieden wird. Dazu wird mittels Objektserialisierung eine Abbildung von der Abstraktionsebene der Ereignisse auf die Ebene der Datenübermittlung (über die TCP/IP Protokollschicht) definiert.

Die für den Anwendungsentwickler interessante Programmierschnittstelle (API) bietet Methoden zum Verbindungsmanagement, Erzeugen und Löschen von Objekten im gemeinsamen Informationsraum und Interessensregistrierung beim Ereignisdienst. Außerdem wird das Versenden von Ereignissen sowie die lokale Behandlung eingehender Ereignisse unterstützt. Der Zugriff auf Objekte aus dem gemeinsamen Informationsraum erfolgt über lokale *Proxy-Objekte*. Methodenaufrufe auf diesen Objekten werden durch einen Objektmanager transparent an den MetaWeb Server übermittelt. Außerdem werden Benachrichtigungen über Zustandsänderungen an Objekten automatisch auf die lokalen Objekte übertragen.

Der im vorherigen Abschnitt beschriebene Monitor des BSCW Systems wurde so implementiert, daß zunächst das BSCW System um einen Ereignisagenten erweitert wurde, der Ereignisse aus dem BSCW System an das MetaWeb System weiterleitet. Dazu wird zu jedem Arbeitsbereich eine Konferenz verwaltet, an der alle Teilnehmer eines Arbeitsbereiches zwecks Benachrichtigung über Aktivitäten teilnehmen können. Dabei findet eine Zugriffskontrolle über das Authentifizierungsverfahren des BSCW Systems statt.

Ereignisagent und Monitor sind als MetaWeb Anwendungen in Java implementiert (s. Abbildung 5). Während Ereignisagent und MetaWeb Server auf Seite des BSCW Systems ablaufen, wird der Monitor als Java Applet auf Seite des Benutzers vom Web Client gestartet und ausgeführt. Beim Start des Monitors aus dem BSCW System erhält dieser als Parameter alle benötigten Informationen über den vertretenden Benutzer (Name, Kennung, Passwort etc.), um sich beim MetaWeb Server anzumelden und an den entsprechenden Konferenzen teilnehmen zu können.

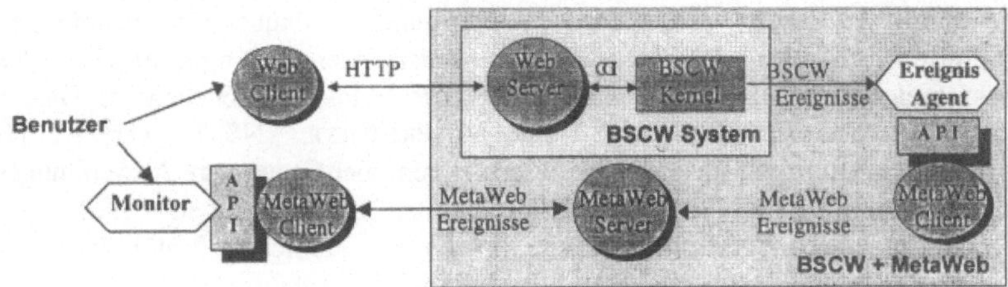

Abbildung 5: Ereignisagent und Monitor als MetaWeb Anwendungen

Eine weitere Anwendung, die mit dem MetaWeb System realisiert wurde, ist ein Demonstrator, der das gemeinsame Betrachten von WWW Seiten ermöglicht und so eine allgemeine Awareness im Web realisiert. (Weitere Informationen hierzu und zu MetaWeb[3] findet man in [13]).

5 Verwandte Arbeiten

Das Ziel, die gleichzeitige Zusammenarbeit mehrerer Benutzer zu unterstützen ist nicht neu und dementsprechend existieren eine Vielzahl von Anwendungen wie z.B. Mehrbenutzer-editoren oder Konferenzsysteme. Da für diese jedoch jedesmal erneut eine gewisse Grundfunktionalität entwickelt werden muß, entstand die Idee der allgemeinen Entwicklungs-unterstützung für synchrone CSCW Systeme. Daraus resultierten verschiedene *groupware toolkits* (Entwicklungsumgebungen für Groupware Anwendungen) wie z.B. RENDEZVOUS [9] oder GROUPKIT [11].

Die starke Verbreitung und Akzeptanz des WWW und die damit verbundene wachsende Bedeutung für die Unterstützung kooperativer Arbeit führten zu einer verstärkten Nutzung des WWW im Kontext von Kooperationssystemen. Die daraus entstandenen Systeme wie ALTAVISTA FORUM[4] oder WEB4GROUPS[5] konzentrieren sich allerdings auf die Unterstützung

---

[3] MetaWeb Home Page - http://bscw.gmd.de/MetaWeb/

[4] AltaVista Forum Home Page - http://altavista.software.digital.com/forum/index.htm

asynchroner Kooperation. Es zeigt sich, daß die einfache Web Technologie für die synchrone Kooperation nur bedingt geeignet ist, während der Einsatz der Java Technologie zu weiteren Systemen wie GROCO [14] oder COWEB [7] führt, die diese Funktionalität auch im Web zur Verfügung stellen.

Die Erweiterung des WWW erfolgt dabei in der Regel durch eine parallele Client/Server Architektur (wie in MetaWeb). Zur Grundfunktionalität dieser Systeme gehört ebenfalls die Verwaltung dynamischer Gruppen (Konferenzen) und teilweise - wie z.B. in CORONA [12] - auch die Verwaltung gemeinsamen Zustands durch verteilte Objekte. Es zeigt sich also, daß diesen Systemen zur Entwicklungsunterstützung eine gewisse Funktionalität gemein ist, die in jedem System neu realisiert wird (ein Umstand, der gerade durch diese Systeme verhindert werden soll).

Ein weiteres Problem stellt dabei die Geschlossenheit dieser Systeme dar, die auf eine Programmiersprache (meist Java) und ein eigenes Protokoll fixiert sind. Will man hingegen Interoperabilität zwischen verschiedenen Anwendungen erlauben, so benötigt man ein gemeinsames Protokoll und Schnittstellen zu verschiedenen Programmiersprachen bzw. Betriebssystemen (wie es im WWW durch das HTTP Protokoll geschehen ist). Dieser Ansatz wird beim NOTIFICATION SERVICE TRANSFER PROTOCOL (NSTP) verfolgt [10], ein Benachrichtigungsprotokoll für die Entwicklung synchroner Groupware Anwendungen. Es ist allerdings fraglich, ob sich NSTP als einheitlicher Standard durchsetzen wird. Inzwischen ist mit dem RENDEZVOUS PROTOCOL (RVP) bereits ein ähnliches Protokoll als Standard vorgeschlagen worden [3].

Betrachtet man die Integration synchroner und asynchroner Kooperationsunterstützung in einer WWW basierten Anwendung, so ist hier (neben dem BSCW System) noch ORBIT [8] zu erwähnen, das ebenfalls Arbeitsbereiche (sog. *locales*) bietet und das Bewußtsein über die Präsenz anderer Mitarbeiter in einem Arbeitsbereich unterstützt. Eine erste Version des Orbit Systems basiert auf dem BSCW System und bietet - ebenfalls mittels Java Applets - eine Gruppenwahrnehmung in gemeinsamen Arbeitsbereichen[6].

6   Zusammenfassung und Ausblick

Wir haben aufgezeigt, wie das primär auf asynchrone Kooperation ausgerichtete BSCW Shared Workspace System - ein Web basiertes Groupware System - um Komponenten zur Unterstützung von synchroner Kommunikation erweitert wurde. Die Erweiterung erfolgte dabei mit Hilfe des MetaWeb Systems, einer Entwicklungsumgebung zur Konstruktion synchroner Kooperationswerkzeuge. Bei der Entwicklung von MetaWeb wurde Java als Basis genommen. Damit konnte das gewünschte Ziel der Unterstützung plattformübergreifender Kommunikation erreicht werden, denn die heute üblichen Browser können Java Applets verarbeiten.

Allerdings bereitet der Einsatz von Java Applets in der Praxis Probleme im Zusammenhang mit der Größe des zu übertragenden Programmcodes. Komplexe Applets wie etwa der oben beschriebene BSCW Monitor, die zudem meist noch auf verschiedenen Klassenbibliotheken

---

[5] Web4Groups Home Page - http://www.web4groups.at/

[6] OrbitLite Home Page - http://www.dstc.edu.au/wOrlds/OrbitLite/

basieren (hier: MetaWeb und IFC - Internet Foundation Classes), erreichen leicht die Größe eines Megabyte und mehr. Da der gesamte Java Bytecode sowie die Programmresourcen bei jedem Start über das Netz übertragen werden müssen, können sich für die Benutzer insbesondere bei langsameren Netzverbindungen unzumutbare Wartezeiten ergeben. Daher haben wir auch das BSCW Monitor Applet derzeit (November 1997) noch nicht auf dem offiziellen BSCW Server der GMD zur allgemeinen Verfügung gestellt.

Abhilfe schaffen hier neuere Ansätze, die eine lokale Speicherung des Programmcodes erlauben. So unterstützen neuere Browser wie Microsofts Internet Explorer[7] und Netscapes Communicator[8] die lokale Installation von Java Klassen, wobei allerdings eine Zertifizierung (digitale Unterschrift) des Programmcodes durch den Anbieter erforderlich ist, damit die eindeutige Herkunft des Codes garantiert ist. Dieses Verfahren soll auch für die Erweiterung des BSCW Systems zum Einsatz kommen, so daß (bei vorhandenem Zertifikat) der Monitor allen Benutzern zur Verfügung gestellt werden kann.

Eine weitere Möglichkeit für die Verteilung und lokale Installation von Softwarekomponenten bietet Marimbas Castanet[9] Technologie. Hier wählt der Benutzer mittels einer lokalen Anwendung (*Tuner*) Software über sog. Kanäle (*Channel*) von einem Anbieter (*Transmitter*) aus. Die Software (z.B. ein Java Applet) wird zum Benutzer übertragen, dort installiert und bei Bedarf (neue Version) inkrementell aktualisiert. Obwohl dieser Ansatz die Installation einer weiteren Anwendung beim Benutzer erfordert, bietet er durch die automatische Aktualisierung wesentliche Erleichterungen bei der Verteilung neuer Versionen und wird daher auch für einen zukünftigen Einsatz im Kontext des BSCW Systems untersucht.

Danksagung

Das BSCW Projekt wird teilweise über das CoopWWW Projekt durch Fördermittel von der Kommission der Europäischen Union im Rahmen des Telematics Applications Programme finanziert. CoopWWW begann im Januar 1996 und hat eine Laufzeit von 24 Monaten. Während dieser Zeit haben mehrere Personen an der Entwicklung des BSCW Systems mitgewirkt, denen an dieser Stelle gedankt werden soll: Richard Bentley, Uwe Busbach, Elke Hinrichs, Thilo Horstmann, David Kerr, Rudolf Ruland, Klaas Sikkel, Jonathan Trevor und Gerd Woetzel.

Literaturverzeichnis

[1] Arnold, K., Gosling J. (1996), "The Java Programming Language", Addison-Wesley 1996

[2] Bentley, R., Appelt, W., Busbach. U., Hinrichs, E., Kerr, D., Sikkel, S., Trevor, J., Woetzel, G. (1997) "Basic Support for Cooperative Work on the World Wide Web" in International Journal of Human-Computer Studies 46 (6), Special issue on Innovative Applications of the World Wide Web, S. 827-846, online verfügbar unter http://bscw.gmd.de/Papers/IJHCS/

---

[7] Microsoft Authenticode - http://www.microsoft.com/ie/ie40/features/sec-authenticode.htm

[8] Netscape Codesigning - http://developer.netscape.com/library/documentation/signedobj/trust/index.htm

[9] Castanet Overview - http://www.marimba.com/doc/general/2.0/introducing/introducing.html

[3]   Calsyn, M. (1997) "RVP - Rendezvous Protocol", Internet Draft, submitted to IETF,
      online verfügbar unter ftp://ietf.org/internet-drafts/draft-calsyn-rvp-00.txt

[4]   Dix, A. (1996) "Challenges and Perspectives for Cooperative Work on the Web", Proc. 5th ERCIM/W4G
      Workshop, GMD Forschungszentrum Informationstechnik, Sankt Augustin, Arbeitspapiere der GMD Nr.
      984, Februar, 1996, S. 143-157, online verfügbar unter
      http://orgwis.gmd.de/projects/W4G/proceedings/challenges.html

[5]   Dourish, P., Bellotti, V. (1992) "Awareness and Coordination in Shared Workspaces", Proc. Conference on
      Computer Supported Cooperative Work, 31.Oktober - 4.November 1992, Toronto, Canada, S. 25-38

[6]   Ellis C.A., Gibbs S.J., Rein, G.L. (1991) "Groupware: Some issues and experiences". Communications of
      the ACM, vol.34, no.1, 1991, S. 38-58

[7]   Jacobs, S., Gebhardt, M., Kethers, S., Rzasa, W. (1996) "Filling HTML forms simultaneously: CoWeb -
      architecture and functionality", Proc. 5th International World Wide Web Conference WWW'96, May 6-10,
      Paris, 1996, S. 1385-1395

[8]   Mansfield, T., Kaplan, S., Fitzpatrick, G., Phelps, T., Fitzpatrick, M., Taylor, R. (1997) "Evolving Orbit: a
      progress report on building locales", Proc. International ACM SIGGROUP Conference on Supporting
      Group Work, 16.-19. November 1997, Phoenix Arizona, USA, S. 241- 250

[9]   Patterson, J.F., Hill, R.D., Rohall, S.L. (1990) "Rendezvous: An Architecture for Synchronous Multi-User
      Applications", Proc. Conference on Computer Supported Cooperative Work CSCW'90, October 7-10, Los
      Angeles, USA, ACM Press, 1990, S. 317-328

[10] Patterson, J.F., Day, M., Kucan, J. (1996) "Notification Servers for Synchronous Groupware", Proc.
      Conference on Computer Supported Cooperative Work CSCW'96, November 16-20, Boston, USA, ACM
      Press, 1996, S. 122-139

[11] Roseman, M., Greenberg, S. (1992) "GroupKit: A Groupware Toolkit for Building Real-Time
      Conferencing Applications", Proc. Conference on Computer Supported Cooperative Work CSCW'92, Oct.
      31- Nov 4, Toronto, Canada, ACM Press, 1992, S. 43-50

[12] Shim H., Hall R., Prakash A., Jahanian, F. (1997) "Providing Flexible Services for Managing Shared State
      in Collaborative Systems", Proc. European Conference on Computer Supported Cooperative Work, Sept.
      1997, Lancaster, Kluwer Academic Publishers, S.237-252

[13] Trevor, J., Koch, T., Woetzel, G. (1997) "MetaWeb: Bringing synchronous groupware to the World Wide
      Web", Proc. European Conference on Computer Supported Cooperative Work, Sept. 1997, Lancaster,
      Kluwer Academic Publishers, S.65-80

[14] Walther, M. (1996) "Supporting Development of Synchronous Collaboration Tools on the Web with
      GroCo", Proc. 5th ERCIM/W4G Workshop, GMD Forschungszentrum Informationstechnik, Sankt
      Augustin, Germany. Arbeitspapiere der GMD Nr. 984, February, 1996, S. 81-89

Adressen der Autoren

Dr. Wolfgang Appelt
GMD - Forschungszentrum Informationstechnik
FIT - Institut für Angewandte Informationstechnik
Forschungsbereich Kooperationssysteme
Schloß Birlinghoven, D-53754 St. Augustin
Email: wolfgang.appelt@gmd.de

Thomas Koch
GMD - Forschungszentrum Informationstechnik
FIT - Institut für Angewandte Informationstechnik
Forschungsbereich Kooperationssysteme
Schloß Birlinghoven, D-53754 St. Augustin
Email: thomas.koch@gmd.de

Social Construction of Knowledge

Angi Voss, Thomas Kreifelts

GMD – Forschungszentrum Informationstechnik
Institut für Angewandte Informationstechnologie (FIT)
53754 Sankt Augustin
angi.voss@gmd.de

Abstract: The term "social construction of knowledge" is introduced and a system is proposed that facilitates this process. Knowledge is based on information that may have to be retrieved from vast, dynamic and unknown information spaces. The goal of knowledge construction may be vague, such that the tasks to achieve it evolve incrementally as information is gathered and insights increase. The collaborative construction of knowledge introduces further challenges and stresses aspects such as validity, trust, relevance, and popularity of knowledge. A combination of groupware, text and data mining tools, software agents and a knowledge building component is proposed to address all these issues involved.

1 Objectives

The work reported here is embedded in a larger research program at GMD-FIT, called "The Social Web". This program is motivated by our belief that the Internet has the potential of being transformed into a comprehensive social medium. By providing a suitable infrastructure, three socially relevant objectives shall be achieved: inventing new forms of culture that people can co-enjoy, reducing isolation, and facilitating knowledge exchange and sharing. It is this last objective that we will address more closely in this paper. For another aspect of the Social Web program, see [Mark & Voss 97].

Diverse interest groups will come to appreciate the Web as a means to achieve their goals that adds new qualities to their interaction. However, the creation of truly useful work environments on the Web raises several challenges: How to design shared virtual spaces and how to represent persons in them so that self-organization and the development of norms and conventions is facilitated? How to collaboratively construct and maintain knowledge, and how to support these processes and keep them alive in a virtual environment? How to increase the reliability of information and people's trust in the shared knowledge?

The Internet presents a vast space of potentially useful information. However, as more information reaches consumers at work or at home, the need for efficient means of creating structure, context and, finally, knowledge is urgent. For a human being to *have knowledge*, as opposed to *being informed*, is to achieve insight enough to act.

Knowledge is rarely developed in isolation. Rather, it is the result of a value adding process where a flow of information originating from human beings is analyzed in explicit or implicit social interaction. We call this process *social construction of knowledge*.

The objective of our work is to develop and apply state-of-the-art computer-mediated communication technology in support of the social construction of knowledge in open or closed groups of collaborating individuals that share a common, but potentially vague, goal.

We explicitly address the challenges of *unknown and dynamic information spaces*, which may be expanding rapidly and contain information of unknown quality, and which may change over time and in unpredictable ways. Such spaces form an unknown territory with no or little a priori knowledge to guide its exploration. Rather, the knowledge has to be extracted dynamically and co-evolves with the information acquired. Even the goal of exploration may be vague so that the tasks to achieve it can be specified only with a growing understanding of the subject matter.

Our work also stresses *collaboration*. People are recognized as sources of intangible knowledge. The formation of groups is understood as requiring a good match between people, tasks, and knowledge. Qualities such as trust, subjectivity, relativity, and publicity of knowledge are taken into account. The collaboration of people is supported by facilities for discussion and communication, for raising cooperation awareness, and for explicit coordination.

In the following, we will elaborate our ideas and outline the envisaged system.

2 The Knowledge Construction Process

We intend to build a software prototype, called the Inter3ight system. It aims at knowledge workers who are experienced users of information spaces like the Internet, but not necessarily IT-specialists. The system will support teams of typically 3 to 30 persons pursuing a common goal like the acquisition of knowledge for a new project, or the monitoring of markets for business intelligence tasks. Professional analysts may apply the system to organize and extract knowledge from large bodies of recordings and documentation. In such applications, knowledge construction is essentially an iterative process which incorporates the following activities.

> *Grounding:* finding suitable and trustworthy information sources, and improving the collection and selection of information.

> *Analysis:* finding clusters and regularities, extracting concepts and relations, recognizing changes and trends, common and divergent views, visualizing the information in a "graspable" way.

> *Construction*: assessing documents, describing and relating concepts, organizing information in terms of "knowledge maps", managing different views and consistent modifications.

> *Coordination*: facilitating the communication among the participants, raising their awareness of each other and of their contributions, organizing their collaboration.

Stressing social and pragmatic aspects, rather than formal and cognitive ones, we want to support the following facets of knowledge (see [Tennison 95] for further arguments):

> *Multimedia:* All types of documents that can be handled by today's computers systems should to be taken into account, from textual documents via graphics to audio and video.

Metadata: To increase the value of information, it should be possible to associate chunks of knowledge that add annotations, categories, or ratings to the underlying documents.

Organization: To describe a collection of documents under different perspectives, it should be possible to create knowledge maps that aggregate, organize and access the documents.

Validity: Knowledge may change as information is revised or more information is gained. On the basis of time-stamps the status of documents should be recognized, the validity of knowledge could be inferred, and trends could be exposed.

Relevance and trust: Knowledge may be more or less relevant for particular goals and tasks and in particular contexts. One's trust in a particular piece of knowledge may depend on its source or on other people's opinion. Therefore knowledge should be related to its authors and information sources, and the exchange and combination of individual assessments and views should be facilitated.

Popularity: Pieces of knowledge may differ in their popularity. Unpopular knowledge may be irrelevant, not well accessible, not trusted, or too novel. On the basis of the event history, one could measure the popularity of knowledge and initiate corrective actions, such as discarding or pushing the knowledge to the relevant people.

Intangibility: Beside tangible and externalized knowledge, intangible knowledge that resides in the heads of individuals has been recognized as a major asset. A competence registry could help to find the right experts for a task. Once experts have been identified, a spectrum of communication facilities should be at their disposal.

3 System Description

To support knowledge construction as outlined above, we intend to integrate different and complementary technologies. At a uniform interface the InterSight system will provide a new blend of functionality that should not necessitate an understanding of the technical backgrounds.

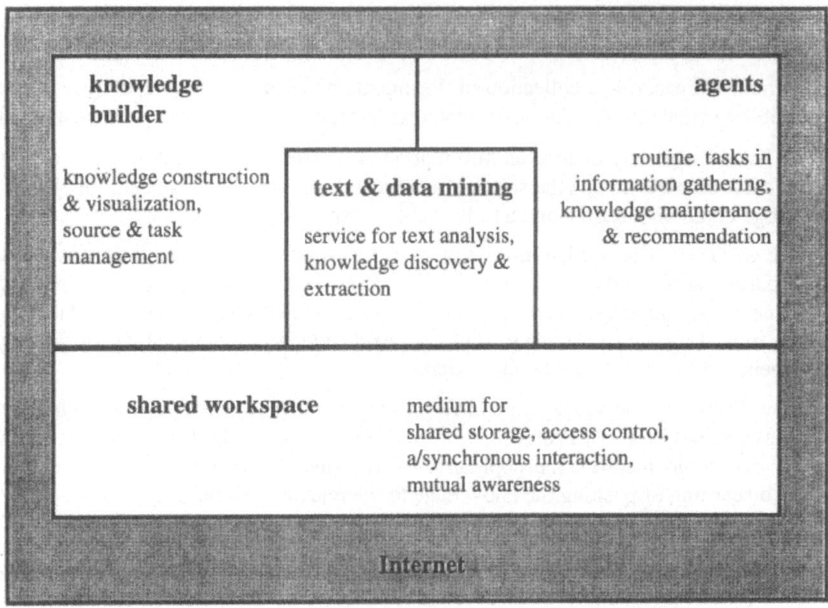

Composition of the InterSight system

A shared workspace offers groupware and interaction functionality. It supports sharing and access control, synchronous and asynchronous communication, and cooperation awareness. It is the "webtop" of the system.

A knowledge builder offers all functionality for interactive knowledge construction. It provides a graphic interface for knowledge visualization and browsing. It allows to assess the information, to categorize, rate, and annotate it. Knowledge maps with concepts and relations can be created picturing the underlying knowledge structure and providing pointers to the information. The specification of tasks is offered as a means to distribute the work among the group.

Mining services support the discovery of knowledge in the document base. The collected documents can incrementally and interactively be clustered and categorized. Concepts and relations can be extracted to elaborate the knowledge maps and to organize the information in the knowledge builder. The contributions of different users can be compared and merged.

Agents are assigned to routine tasks of knowledge construction and maintenance. On behalf of the users they gather and periodically update information from external sources, they apply the mining services to organize the information in knowledge maps and to integrate new information into the maps. The agents collect knowledge about information sources and about the competence of people to recommend them for new tasks.

3.1 BSCW for Shared Workspaces

The shared workspace component handles all groupware and interaction functionality and provides the webtop for the complete system. We will use BSCW, our system for shared workspaces on the Web (http://bscw.gmd.de/) [Bentley et al. 97]. BSCW has a diverse and active community of some ten thousand users from industry and research which, in the past, has shaped BSCW through its fast and rich feedback.

Central to BSCW is the concept of a shared workspace which the members of a group establish for organizing and coordinating their work. A shared workspace may contain documents of different types, links to other Web pages or FTP sites, threaded discussions, member contact information and more. Workspace members can set access rights to control the visibility of this information and the operations which can be performed by others. An event service provides basic awareness. BSCW is being extended by services for launching various Internet-based synchronous and asynchronous conferencing tools, advanced awareness facilities, and a Java-based user interface.

For its integration into the InterSight system, BSCW's object types have to be extended to include new kinds of documents for the objects manipulated in the knowledge builder: tasks, information sources, competence registry, knowledge maps, and background knowledge. The documents may have new kinds of attributes for metadata like ratings and categories.

3.2 Knowledge Builder

The knowledge builder will support the users in defining and conducting the knowledge construction process. It provides access to services of the agent and mining components. The knowledge builder is activated through the shared workspace and uses its support for synchronous and asynchronous collaboration, access control, and awareness. The knowledge builder will cover knowledge creation as an iterative process as described above.

Knowledge construction tasks: Many organizations have their own preferred way of working, some more detailed, prescriptive and sophisticated, others more general and liberal. The knowledge builder will offer decomposable tasks and a planning matrix to match objectives with information requirements and available sources. Templates and methods will be available to guide users with different experiences, similar to wizards in many office suite software packages.

Information retrieval: The knowledge builder allows to configure agents that automate information retrieval and monitoring. The agents may access search engines or controlled Internet sources like Reuters and Knight-Ridders. To find suitable information sources, the knowledge builder will allow users to add descriptive information about a particular source's value in terms of actuality, trustworthiness, timeliness etc. Agents will update the information about a source continuously based on user feedback.

Analysis: The knowledge builder offers text and data mining services to analyze the collected information, to organize it, to extract knowledge from the information, and to differentiate or merge different contributions. Visualization methods support navigation through organized document collections, queries, knowledge maps, and the background knowledge. Users can directly operate on the knowledge presentations, and modify them to reflect their view.

Knowledge construction and organization: Knowledge is constructed both manually and semi-automatically through structures suggested by the text mining service, which the users may accept, refuse, or modify. Elementary chunks of knowledge can be attached to individual documents. This can be categories, ratings or annotations. Complex knowledge maps can be formed introducing and relating concepts, and linking them to the underlying documents, thus imposing meaningful patterns on the information. Users can configure agents that update and maintain this knowledge.

Dissemination and presentation: The results from a knowledge construction task have to be communicated to other users of the system, within and/or outside of the specific task. There can be special knowledge maps for just-in-time presentations which the agents keep up to date and the knowledge builder will include version control of knowledge structures.

An additional function of the knowledge builder is a *registry of competence*. This function helps in identifying individuals that have shown to be knowledgeable in a particular subject area. The "ranking" of these individuals can be specified manually in the knowledge base, or be derived from their activities and work in the shared workspace.

3.3 Mining Services

The mining component provides services to the knowledge builder and the agents. It offers a selection of complementary methods from pre-processing to maintenance. The methods will use available linguistic and task-specific background knowledge. Task-specific knowledge may come from imported ontologies or previous tasks. Linguistic knowledge will be treated in a modular fashion so that English can be exchanged by other languages.

Pre-processing: Efficient pre-processing methods have to be implemented to extract terms from text documents or from metadata of multimedia documents. The resulting set-based representation, where each document is represented by a set of weighted terms, is processed to derive categorizations and term or concept hierarchies.

Clustering partitions a document collection into clusters of similar documents. Each cluster represents a set of documents with common characteristics (e.g. terms). Hierarchical clustering approaches with optional user specified biases provide an insight into the structure of the collection. Each cluster is summarized by information on the terms that are relevant for its documents.

Comparative distribution analysis compares the distribution of terms in different sub-collections of documents. Thus the user can gain information on the differences and similarities between the subcollections and on the significance of the terms in each subcollection. Distribution analysis results will be presented by tables and graphs.

Association discovery methods provide information on the co-occurrence and implications between terms. Association discovery is constrained by various constraint types to avoid the user from drowning in a huge amount of results.

Knowledge maintenance and revision methods support incremental updates and combinations of knowledge maps. These methods combine previous knowledge with new one to detect any significant changes in the knowledge, and they combine knowledge from different sources to detect and possibly resolve contradictions.

Profiles will be computed with standard statistical methods to characterize information sources, user competence or interest, and document ratings.

Mining services will be built as extensions of existing technology: stemmers and thesauri, text mining and visualization algorithms as in the Document Explorer [Feldman & Klösgen 97], structural similarity measures for clustering combined with Bayesian methods, data mining algorithms from GMD's Kepler system [Wrobel et al. 96], and knowledge revision algorithms from GMD's Mobal™ system [Wrobel 96], an integrated inductive logic programming system for knowledge maintenance and revision.

3.4   Agent Services

Agents are long-lived software processes that perform tedious and repetitive activities to achieve a specific goal for their clients. For the InterSight system, agent services will be developed for gathering information, updating and maintaining knowledge on behalf of users and groups. For these purposes, the agents will employ mining services and other system functions.

Competence registry: The agents use the mining service to gather and update evidence about the competence of individuals as can be found in their contributions to previous tasks. They correspondingly update the competence registry in the knowledge builder, which can be used to find the best experts for a given task.

Recommendation of information sources: Using information from previous tasks, the agents gather and update assessments of information sources in the knowledge builder so that the most suitable information sources can be recommended for a given task. Given a specification of the desired information, the agents search the Web for new relevant information sources.

Information gathering service: In the context of a task, the agents query the specified information sources in order to periodically obtain new information or to update previously collected information.

Information organization: When the users have constructed a knowledge map with concepts and relations, the agents can use mining services and background knowledge to derive queries. They contact information sources to gather information and link it to the user's knowledge map.

Knowledge map maintenance: In a similar way, the agents can incrementally update a knowledge map. The agents use the mining services to detect changes in the information and possibly affected knowledge, and suggest changes to the knowledge map.

System knowledge maintenance: On behalf of the knowledge builder and by using the mining services, the agents can assess the popularity of knowledge. They can identify irrelevant knowledge, push ignored but important knowledge to the right people, and perform any house keeping functions that, if applied regularly, keep the knowledge up to date.

For each service, special kinds of agents have to be built. We have implemented a demonstrator agent system, called LiveMarks, that supports groups of people in sharing, assessing, recommending and updating bookmark collections. These agents and their underlying agent infrastructure, the SOaP system, will be incorporated into the InterSight system. The SOaP agent infrastructure simplifies the building of lightweight, secure, open, robust, and distributed communicating agents.

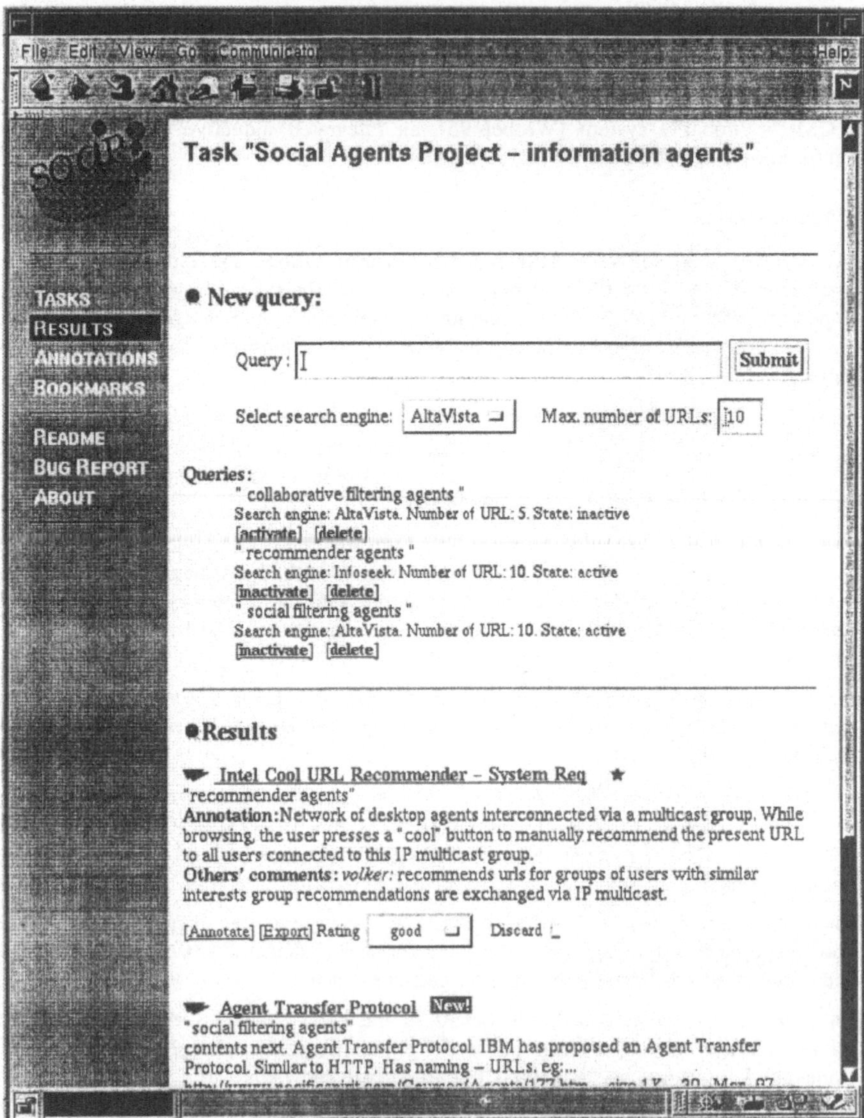

A task page in LiveMarks

LiveMarks can be viewed as providing dynamic, or live, bookmark pages [Voss & Kreifelts 97]. A user agent maintains a list of personal tasks for each user. Each task is maintained by task agent. A task consists of several active or inactive queries. An active query is periodically repeated to gather new information. Secondly, a task contains a list of results which the user may rate, annotate, or discard. Thirdly, a task may identify a group for which the task results are evaluated. The task agent gets results by forwarding the queries to a search agent, which contacts selected search engines, and to a recommender agent. The recommender agent collects results from all users with their ratings, annotations, and group membership. The recommender agent answers a query by retrieving results from group members with similar queries and with high ratings.

3.5   The InterSight System

As we envisage continued rapid technological development in the area of all component technologies during the next years, it is important to design the InterSight system for exchange. The figure below shows the layered and modular architecture of the system.

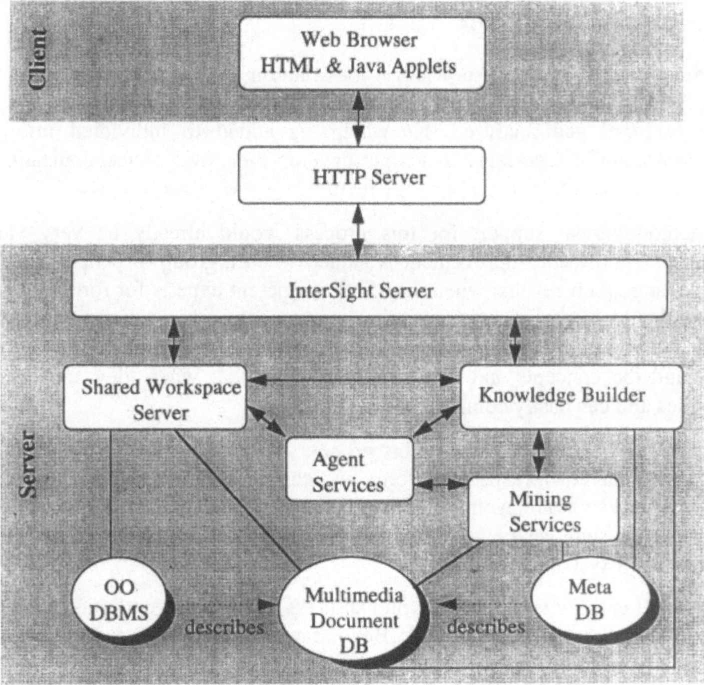

Architecture of the InterSight system

User interface: The InterSight system will be implemented using Java on both client (applets) and server (servlets). In addition, some parts will be implemented as HTML generators. As a consequence, the system's user interface can easily be replaced or reconfigured.

Databases: The InterSight system will use three different kinds of databases, a metadata base that holds the added structures as metadata, a multimedia document database, and an object-oriented database that holds the descriptions of objects for the shared workspace component. The metadata base will be a relational database. The relational database technology is well standardized which allows us to use database products from different vendors, if needed. Although the multimedia document database market is not as well standardized, we will design the InterSight system to be independent of the choice of the document database.

InterSight server: The InterSight server wraps the components and provides a uniform point of access from outside. The shared workspace platform invokes the knowledge builder when one of its objects is accessed. Through the knowledge builder, agent and mining services are accessible. The mining services operate on the document and metadata base, while the agents access any data only indirectly through the other components.

4 Conclusion and Next Steps

We have presented our view of knowledge construction as an iterative process. Gathering information from selected sources leads to a set of disorganized information pieces, which have to be explored and analyzed. Knowledge is added to individual information by annotating, rating, and categorizing it. People develop their own conceptualization to index and map the information, to structure and restructure it as their insights grow.

Complete methodological support for this process would already be very valuable for individual users. Significant improvement is gained when a group of people can collaborate and easily exchange their results: when they find competent experts for forming teams; when they can share their assessments of information sources; when they can decompose their tasks and distribute information gathering; when they can compare and combine their assessment of information, and the concepts and structures they extracted; when they are aware of each other's activities and can easily communicate and interact.

We have outlined a system to enhance this process of socially constructing knowledge. Its corner stones are a shared workspace platform, an interactive knowledge builder, text and data mining services, and software agents. To implement the InterSight system we are proceeding bottom up. As a first step, we are going to integrate the BSCW shared workspace system with the LiveMarks agent system.

Within BSCW, all activity is initiated by human users, there is currently no concept for automating repetitive or time-consuming work. Beyond answering queries with recommendations of bookmarks, LiveMarks lacks any functionality for explicitly sharing complete information gathering tasks contexts with other users. Merging BSCW with LiveMarks will present substantial progress:

BSCW users can issue queries that are automatically repeated. The results of their queries are filtered in the context of a folder so that a bookmark does not show up again once it has been discarded or rated. BSCW users obtain recommendations of bookmarks based on collaborative filtering.

LiveMarks users can deposit and use their bookmarks in the context where they are relevant, and they can share them with others. They can distinguish general annotations of a bookmark, which correspond to a description in BSCW, from opinions and organizational remarks, for which they can use BSCW's notes and discussion facilities. BSCW's awareness service indicates when new bookmarks have been received or existing ones got a new annotation.

Acknowledgments

We would like to thank Ulf Wingstedt for his contributions to the InterSight system architecture and the knowledge builder, Willi Klösgen for his contributions to the mining services, and our colleagues from the BSCW and agents projects at GMD-FIT for their discussions about the other components.

References

Bentley, R.; Appelt, W.; Busbach, U.; Hinrichs, E.; Kerr, D.; Sikkel, K.; Trevor, J.: Basic support for cooperative work on the World-Wide Web. In: International Journal of Human Computer Studies: Special Issue on Innovative Applications of the World-Wide Web, **46**, 6 (1997), 827-846.

Feldman, R.; Klösgen, W.: Document Explorer: Discovering knowledge in document collections. In: Ras, Z.; Michalewicz, M. (eds.): Foundations of Intelligent Systems. Lecture Notes in Artificial Intelligence. Berlin: Springer, 1997.

Mark, G.; Voss, A.: Attractivity in virtual environments: Getting personal with your agent. In: Dautenhahn, K. (ed.): Proceedings AAAI Fall Symposium 1997: Socially Intelligent Agents, (Technical Report FS-97-02), Nov. 8-10, 1997 Cambridge, MA. Menlo Park, CA: AAAI Press, 1997, pp. 81-86.

Tennison, J.: Living Documents: Adding collaboration to information systems. Presented at the Postgraduate Conference, Department of Psychology, University of Nottingham, June 8-9, 1995.
http://www.psyc.nott.ac.uk/aigr/papers/Living-Documents/paper.html

Voss, A.; Kreifelts, T.: SOaP: Social agents providing people with useful information. In: Hayne, S. C.; Prinz, W. (eds.): Proceedings GROUP '97, International ACM SIGGROUP Conference on Supporting Group Work – The Integration Challenge, Nov. 16-19, Phoenix, AZ. New York, NY: ACM, 1997, pp. 291-298.

Wrobel, S.; Wettschereck, D.; Sommer, E.; Emde, W.: Extensibility in data mining systems. In: Simoudis, E.; Han, J.; Fayyad, U. (eds.): Proceedings 2nd International Conference on Knowledge Discovery and Data Mining. Menlo Park, CA: AAAI Press, 1996.

Wrobel, S.: Systems with domain competence: Knowledge acquisition assistance. In: Hoschka, P. (ed.): Computers as Assistants – A New Generation of Support Systems. Hillsdale, NJ: Lawrence Erlbaum, 1996, pp. 81-95.

Benutzerraum und Dokumentenraum Nachbarschaft im WWW

Klaus H. Wolf und Konrad Froitzheim
Abteilung Verteilte Systeme, Universität Ulm

Zusammenfassung:
Das World Wide Web ist der vielseitigste Dienst im Internet. Das Web bietet Unterhaltung, Nachrichten, Bilder, Software, Reisen, Bücher und viele andere Dinge, die man im täglichen Leben braucht. Aber ein Aspekt, der in der realen Welt eine wichtige Rolle spielt, fehlt bisher: die Möglichkeit andere Menschen zu treffen. Zwar gibt es sogenannte Virtual-Neighborhoods in Form von Chat-Räumen. Die Möglichkeit Menschen zu treffen ist aber beschränkt auf sehr wenige, speziell dafür vorgesehene Webseiten. Wir beschreiben ein Modell und ein Softwaresystem, das aus dem gesamten Web einen Treffpunkt macht. Es ermöglicht den Benutzern des Webs, sich auf jeder Seite zu treffen. Das System basiert auf dem Konzept der dynamischen Nachbarschaft (dynamic neighborhood). Mit Hilfe der dynamischen Nachbarschaft werden Benutzer des Web gegenseitig sichtbar gemacht. Die so sichtbaren Benutzer können dann Kommunikationskanäle öffnen, um miteinander in Ton und Bild zu sprechen.

1 Die Einsamkeit des Web-Surfers

Für Viele ist das Internet zu einem unentbehrlichen Hilfsmittel geworden. Das World Wide Web (hier nur als Web bezeichnet) ist bei weitem der häufigst benutzte Internetdienst für die Suche nach Informationen. Das Web basiert auf asynchroner Kommunikation, insbesondere der Verteilung von Dokumenten verschiedener Typen. Synchrone Kommunikation zwischen Personen ist kein Dienst des Web, kann aber auf einfache Weise integriert werden durch sogenannte Protocol-Handler, Plug-Ins, externe Programme (Helper Applications) oder andere Erweiterungsmöglichkeiten von Web-Klienten.

Millionen Menschen sind zur gleichen Zeit online im Web, oftmals sogar Hunderte, manchmal Tausende auf dem gleichen Web-Server oder der gleichen Web-Site. Aber Benutzer im Web können sich normalerweise nicht gegenseitig sehen. Jeder ist für sich allein unterwegs. Außer Verzögerungen und längeren Wartezeiten zu bestimmten Tageszeiten gibt keine Anzeichen, daß noch andere Menschen gleichzeitig online sind. Surfen im Web gleicht einem Stadtbummel in leeren Straßen. Es ist, als ob die Geschäfte und Bibliotheken voll von Waren sind, aber in den virtuellen Kaufhäusern ist nicht einmal ein Verkäufer zu finden. Ob dies ein Vorteil oder Nachteil des Web im Vergleich zur realen Welt ist, soll dem Leser überlassen bleiben. Tatsächlich sind wir es aber gewohnt, Menschen zu treffen. Darüber hinaus ist es zweifelhaft, ob es sich die Betreiber von virtuellen Geschäften und Kaufhäusern auf Dauer leisten können, potentielle Kunden nicht persönlich anzusprechen.

Das Web erscheint leblos und statisch. Dies ist aber nicht die Wirklichkeit. Tatsächlich gibt es viele Menschen im Web. Manche springen schnell von Seite zu Seite, andere gehen langsam und lesen die angebotenen Inhalte genau. Menschen bewegen sich im Web, um zu arbeiten, Informationen zu suchen oder einfach um sich die Zeit zu vertreiben. Es gibt sogar unbemannte Vehikel, genannt Robots, auf den Straßen des Web, die automatisch Informationen sammeln. Das Web ist also erfüllt mit Leben. Leider ist dieses Leben unsichtbar für den normalen Benutzer.

Diese Analogie mit der realen Welt zeigt, daß auf dem Web eine wichtige Dimension fehlt, nämlich die Benutzer. Werden die Benutzer hinzugefügt, dann wird das World Wide Web zum richtigen Cyberspace.

1.1 Cyber-Nachbarschaft

1.1.1 Dynamische Verzeichnisdienste

Dynamische Verzeichnisdienste (Dynamic Directory Services, DDS) sind ein einfacher Ansatz um Web-Benutzern mitzuteilen, daß Andere gleichzeitig online sind (siehe [1][2]). Zur Verwendung dynamischer Verzeichnisdienste erzeugen die Benutzer Listen von Freunden und Bekannten (sogenannte Buddy-Lists) und schicken diese an den DDS-Server. Der Server benachrichtigt dann jeden registrierten Benutzer, sobald jemand von der zugehörigen Buddy-Liste den Web-Browser startet. Dynamische Verzeichnisdienste geben aber nur einen flüchtigen Eindruck von der lebhaften Natur des Web. Sie sind nicht flexibel genug, um Zufallstreffen zu ermöglichen, die einfach nur deshalb zustande kommen, weil sich zwei Menschen gleichzeitig für dieselben Themen interessieren und auf den gleichen Web-Seiten surfen. Deshalb kann man auf diese Art niemals neue Partner oder Freunde im Cyberspace treffen. Außerdem müssen alle Teilnehmer eine spezielle Software, den DDS-Klienten, installieren.

1.1.2 Virtuelle Konferenzräume - Statische Nachbarschaft

Virtuelle Konferenzräume (Static Neighborhoods / Web Communities) bestehen normalerweise aus einer kleinen Zahl fest vorgegebener Web-Seiten. Menschen, die sich auf diesen Seiten bewegen, sind im gleichen virtuellen Konferenzraum (Virtual Meeting Room, VMR). Im Gegensatz zu dynamische Verzeichnisdiensten können sich in VMRs Menschen treffen, die sich vorher noch nicht kannten. Der Benutzer gibt nur den Ort des VMRs an und nicht die Liste der Partner in Buddy-Lists. Die Liste der sichtbaren Partner wird vom VMR zur Laufzeit bestimmt.

Es gibt viele verschiedene Ausführungen von VMRs. Sie reichen von statischen HTML-Seiten über sogenannte Chat-Räume bis zu komplexen Datenbanksystemen. Sogar dynamische Verzeichnisdienste (DDS) können verwendet werden, um virtuelle Konferenzräume zu simulieren. In diesem Fall erzeugt der VMR-Server die Buddy-Lists dynamisch.

Virtuelle Konferenzräume haben, wie reale Konferenzräume, einen festen Platz. Menschen gehen zu diesem Ort und treffen dort Andere, die die gleiche Web-Seite gewählt haben. Alle Personen in einem VMR können sich gegenseitig sehen. Der sichtbare Bereich - die Web-Seiten - ist gleich für alle Teilnehmer. Er ist definiert durch die Konfiguration des virtuellen Konferenzraums. Der sichtbare Bereich ist damit unabhängig von der tatsächlichen Position eines Teilnehmers innerhalb des VMRs.

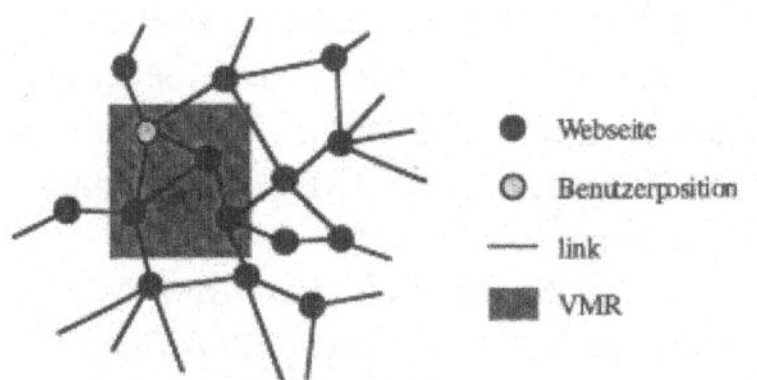

Abbildung 1: Beispiel für einen virtuellen Konferenzraum (Virtual Meeting Room, VMR) aus mehreren Web-Seiten.

Virtuellen Konferenzräume modellieren geschlossene Räume. Dies ist vor allem für solche Aktivitäten sinnvoll, die am besten in geschlossenen Gruppen durchgeführt werden. Beispiele hierfür sind synchrone oder asynchrone gemeinsame Arbeit (CSCW), Besprechungen oder Schulungen bis hin zu Vorlesungen. Geschlossene Räume sind aber nicht geeignet für viele andere Aktivitäten im Web, wie Informationssuche oder Einkaufsbummel.

1.2 Dynamische Nachbarschaft

Das Modell der dynamische Nachbarschaften allgemein in dem Sinne, als es die oben beschriebenen statischen Nachbarschaften, bzw. virtuellen Konferenzräume und die dynamischen Verzeichnisdienste beinhaltet. Wir werden zeigen, daß dynamische Nachbarschaften besser geeignet sind, die dynamische Natur des Web darzustellen. In diesem Kapitel wird der Begriff der dynamische Nachbarschaft eingeführt. Das nächste Kapitel geht genauer auf das von uns verwendete Dienstemodell ein. Dieses Dienstemodell bildet die Grundlage für die in Kapitel 3 beschriebene Implementierung. Im vierten Kapitel werden Ergebnisse vorgestellt und ein Einblick in die aktuelle und zukünftige Arbeit gegeben.

In der realen Welt ist der übersehbare Bereich für jede Person verschieden. Jeder sieht deshalb eine verschiedene Untermenge von Objekten. Überträgt man dies auf das Web, dann muß für jede Person eine eigene (persönliche) Nachbarschaft berechnet werden. Der sichtbare Bereich (die Nachbarschaft) darf nicht nur von der Umgebung - den Web-Seiten - abhängen, sondern muß auch vom Benutzer, seiner genauen Position im Web oder anderen Eigenschaften beeinflußt werden. Während die statische Nachbarschaft nur auf der Konfiguration der Umgebung basiert, bezieht die dynamische Nachbarschaft auch den Benutzer mit ein. In der statischen Nachbarschaft ist der Sichtbarkeit aller Benutzer gleich, da sie durch die Konfiguration des virtuellen Konferenzraums voreingestellt ist. Bei der dynamischen Nachbarschaft wird dagegen der sichtbare Bereich für jeden Benutzer individuell berechnet. Die Berechnung basiert dabei auf Eigenschaften und Konfigurationen des Benutzers, die sich dynamisch ändern können.

Die wichtigste und offensichtlichste Eigenschaft eines Benutzers des Web ist die Position der besuchten Web-Seiten. Wie in der realen Welt bestimmt diese Position den sichtbaren Bereich und damit die Menge der sichtbaren Objekte in der Umgebung. Im Web wird die Umgebung repräsentiert durch Web-Seiten und Verbindungen zwischen den Seiten (Hyper-Links). Die Anordnung von Seiten durch die Verbindungen beeinflußt die sichtbare Nachbarschaft. Wenn Personen sich bewegen, dann führen sie ihre individuelle Nachbarschaft mit sich. Dabei ändert sich die Nachbarschaft entsprechend der Position und den umgebenden Web-Seiten und Hyper-Links. Im einfachsten Fall (Abbildung 2) besteht die dynamisch berechnete Nachbarschaft aus den Web-Seiten, die in der Nähe der jeweils besuchten Seite sind. Gemessen wird die Nähe dabei in Hypertext-Links. Der Benutzer sieht alle benachbarten Seiten und alle Objekte, die mit diesen Seiten assoziiert sind, z.B. andere Benutzer auf diesen Seiten. Bewegt sich ein Benutzer, dann ändert sich die Menge der sichtbaren Objekte. Gleichzeitig sehen andere Benutzer denjenigen, der sich bewegt, in ihrer Nachbarschaft auftauchen und wieder verschwinden.

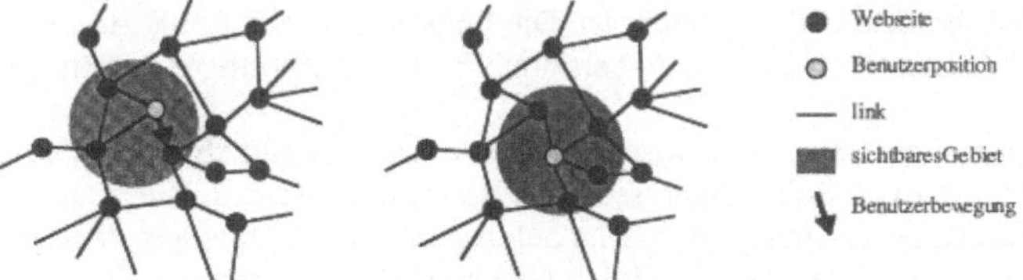

Abbildung 2: Beispiel für eine Person, die sich von einer Web-Seite zur nächsten entlang eines Hyper-Links bewegt. Die Nachbarschaft, d.h. der sichtbare Bereich, ist mit der Person verbunden. Sie bewegt sich mit der Person und ändert sich dabei.

Im Gegensatz zur realen Welt gibt es im Cyberspace mehr Freiheit bei der Definition von Nachbarschaft. Es wurden schon abstrakte Nachbarschaften realisiert, die statt durch Positionen im Web (URLs), durch andere Parameter definiert sind. Ein sehr schönes Beispiel dafür

ist die Wissens- und Interessen-basierte Nachbarschaft des WerWeissWas-Dienstes [3]. Diese
Nachbarschaft ist ebenfalls benutzerorientiert. Sie hängt aber nicht von Positionen im Web ab,
sondern sie ist definiert durch die Ähnlichkeit von Interessensgebieten.
Es ist der Zweck aller Nachbarschaftsmodelle Personen in Beziehung zu setzen. Dies soll auf
eine Art geschehen, die den Nutzen des Internets, bzw. des Web vergrößert. Ein Nachbarschaftsdienst hat deshalb die Aufgabe sinnvolle, d.h. für die Benutzer nützliche, Assoziationen mit anderen Benutzern herzustellen. Um diese Aufgabe zu erfüllen muß ein Nachbarschaftsdienst Informationen über die Benutzer erlangen. Wir verwenden dabei frei verfügbare
Informationen über Bewegungen von Benutzern (durch Spurverfolgung) und Informationen,
die die Benutzer explizit eingeben (z.B. Interessen, wie der WerWeissWas-Dienst).
Das Web - URLs und Links - bildet die Infrastruktur und bietet Inhalte. Dies ist der mit Web-
Klienten sichtbare Teil des Cyberspace. Andere Objekte sind unsichtbar, obwohl sie vorhanden sind. Der Nachbarschaftsdienst wird als zusätzlicher Dienst eingeführt, um - wie ein
Nachtsichtgerät - diese verborgenen Objekte für den Benutzer sichtbar zu machen. Im nächsten Kapitel wird das Modell eines dynamischen Nachbarschaftsdienstes vorgestellt, der dies
leistet.

2 Modell des Nachbarschaftsdienstes

2.1 Orte und Wege im Web

2.1.1 Seiten

Wir glauben, daß die durch das World Wide Web gegebene Infrastruktur ein gutes Fundament
für eine virtuelle Welt darstellt. Adressen und Links im Web basieren auf sogenannten Universal Resource Locators (URL). URLs und ihre Obermenge Universal Resource Names
(URN [4]) repräsentieren Orte im Web. Dies wird vermutlich für längere Zeit (mehrere Jahre)
so bleiben. Obwohl wir hier meistens den Begriff URL verwenden werden, können alle Konzepte gleichermaßen auf URNs angewendet werden, sobald diese in größerem Maßstab im
Web verwendet werden.
Wir betrachten Web-Seiten und alle anderen Arten von Dokumenten, die über das Netz verfügbar sind, als Orte in der virtuellen Welt. Diese virtuellen Orte sind das Gegenstück zu Orten in der realen Welt, wie Räume, Straßenecken und Geschäfte. Personen bewegen sich im
Web zwischen virtuellen Orten über Hypertext-Links. Hypertext-Links sind die Verbindungen zwischen den virtuellen Orten. Sie entsprechen den Straßen und Wegen der realen Welt.
Der einzige Unterschied zu realen Straßen besteht darin, daß es nicht möglich ist, auf den
virtuellen Straßen anzuhalten, bevor das Ziel erreicht ist. Die Funktion der Straße als Aufenthaltsort wird im Web von sogenannten Indexseiten übernommen, die als Zubringer zu den
Inhaltsseiten dienen.
Orte im Web sind nicht notwendigerweise nur statische HTML-Seiten. Es gibt abstrakte Orte,
wie die oben erwähnten, die auf persönlichen Interessen von Benutzern basieren. Diese Orte
sind nicht Teil des Adressenraumes, der durch die URLs aufgespannt wird. Trotzdem können
sie für den Zweck der dynamischen Nachbarschaft durch URLs repräsentiert werden. Ihre
URLs bestehen dann aus der Adresse des Diensteanbieters (z.B. www.wer-weiss-was.de) oder
des Datenbankservers und den Parametern des Benutzers (z.B. einer Zeichenkette kodierter
Interessensbeschreibungen).

2.1.2 Links

Hypertext-Links dienen momentan nur als Verbindungen zwischen Orten (Seiten). Tatsächlich haben sie aber oft noch andere Eigenschaften. Ein Beispiel dafür ist das rel-Attribut von HTML-links. Das rel-Attribut enthält zusätzliche Information über die Art des Links. In unserer Implementierung verwenden wir ein zusätzliches Attribut zum Anker-Tag (HTML <A>-Tag), das über die Länge eines Links Auskunft gibt. Dieses Attribut (genannt "distance") ist optional. Es enthält ein Maß für die Stärke der Verbindung. Orte können durch dieses Attribut weit entfernt sein, obwohl sie durch einen Hypertext-Link direkt verbunden sind. Dies ist sehr nützlich, um Seiten zu trennen, die miteinander verbunden, inhaltlich aber nicht verwandt sind. Beispiele für solche Links gibt es auf sogenannten Hotlist- oder Bookmark-Seiten, die Verweise auf viele, nicht inhaltlich verwandte, Seiten enthalten. Der umgekehrte Fall ist eine Distanz von Null. Durch die Null-Distanz können mehrere Seiten zu einer virtuellen Seite kombiniert werden. Natürlich wird das Distanz-Attribut auf unmodifizierten Web-Seiten (Web-Servern) normalerweise fehlen. Der Standardwert für diesen Fall ist "1". Das Distanz-Attribut enthält entweder einen numerischen Wert in Einheiten von Hypertext-Links (Fließkommawerte) oder eine textuelle Konstante als sogenannte Named-Relationship. Momentan sind die Konstanten "near", "far", und "unrelated" definiert.

Die weitaus meisten Hypertext-Links im Web sind nur unidirektional. Dies ist der Grund für das bekannte "dangling links"-Problem (siehe zum Beispiel [5]). Im Rahmen des Nachbarschaftsdienstes wurde dagegen entschieden, Links zwischen Dokumenten symmetrisch, d.h. bidirektional zu behandeln, da damit eine symmetrische Sichtbarkeit gewährleistet wird, die der alltäglichen Erfahrung entspricht. Diese Entscheidung ist aber nicht Teil des Nachbarschaftsmodells. Sie kann in einer anderen Implementierung des gleichen Dienstes anders ausfallen, ohne die Kompatibilität zu verletzen.

2.1.3 Konferenzräume

Das Distanz-Attribut "unrelated" bedeutet eine komplette Trennung von Orten so, wie reale Wände die Räume eines Gebäudes trennen. Obwohl Personen sich problemlos, wie gewohnt, zwischen Web-Seiten bewegen können, können sie doch nicht sehen, was sich hinter der durch das "unrelated"-Attribut entstandenen Wand befindet. Tatsächlich wird dies realisiert, indem die als "unrelated" markierten Links in HTML durch den Nachbarschaftsdienst ignoriert werden. Da Web-Klienten das Distanz-Attribut ignorieren, wird Web-Browsen über einen solchen Link nicht behindert.

Diese Möglichkeit wird dazu verwendet, um virtuelle Konferenzräume und geschlossene Gruppen zu erzeugen. Virtuelle Konferenzräume werden im Web durch URLs und Web-Seiten repräsentiert. Diese Seiten können durch Hypertext-Links, wie alle anderen Seiten in das Web eingebunden werden, sind jedoch als "unrelated" markiert, und wirken deshalb wie Wände um den Konferenzraum. Für geschlossene Gruppen können bei Bedarf durch Mechanismen des Web (HTTP) zusätzlich auch Zugriffsbeschränkungen eingeführt werden. Man kann sich leicht eine Liste von Konferenzräumen vorstellen, die Hypertext-Links in verschiedene Räume besitzt. Durch das "unrelated"-Attribut ist es nicht möglich in die Räume hineinzusehen. Die Räume sind verschlossen durch Web-eigene Authentifizierungsmechanismen. Unter Umständen könnte man sogar trotzdem den Blick von innen nach außen zulassen. Dies zeigt, daß die Länge von Links nicht notwendigerweise symmetrisch sein muß.

Statische Nachbarschaften fügen sich so nahtlos in das Konzept der dynamischen Nachbarschaften ein. Statische Nachbarschaften sind nur ein Spezialfall der dynamischen Nachbarschaften. Die Web-Seiten sind durch entsprechende Distanz-Attribute so arrangiert, daß alle

Personen in einem Konferenzraum die gleiche Sicht, d.h. den gleichen sichtbaren Bereich haben.

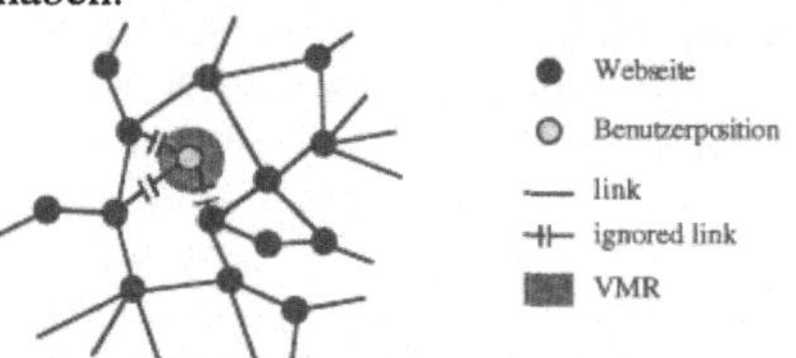

Abbildung 3:   Der sichtbare Bereich ist beschränkt durch Hypertext-Links, die als "unrelated" markiert sind. Die Seite ist für den Nachbarschaftsdienst isoliert vom Web. Isolation bedeutet hier, daß es für Personen auf dieser Seite nicht möglich ist, die Umgebung zu beobachten und umgekehrt.

2.2 Personen und Kommunikation

Die virtuelle Welt besteht nicht nur aus Seiten und Verbindungen. Personen und andere aktive Einheiten, z.B. Robots, agieren in dem Raum, der durch die URLs aufgespannt ist. Wir unterscheiden bisher zwei Aktionsformen. Die erste ist Bewegung durch den virtuellen Raum, wie oben beschrieben. Die zweite Aktionsform ist Kommunikation zwischen den aktiven Einheiten, z.B. Personen oder mobilen Agenten. In der realen Welt sehen wir nicht nur Personen, die sich in der Nähe befinden, sondern wir kommunizieren auch mit ihnen auf verschiedene Art und Weise. Oder wir registrieren Kommunikation, die zwischen anderen Personen stattfindet. Sowohl Personen, als auch Kommunikation bereichern das Web um neue Dimensionen. Während das World Wide Web die Infrastruktur bildet, bzw. das Substrat darstellt, existieren Personen und Kommunikation innerhalb dieser Umgebung. Natürlich sind Personen, die sich im Web bewegen nicht gleichzeitig überall. Das würde der Erfahrung aus der realen Welt widersprechen. Die Präsenz von Personen ist beschränkt auf die betrachteten Web-Seiten (die Position im Web) und die Umgebung dieser Seiten. Die Stärke der Präsenz - oder die Sichtbarkeit durch andere Personen - wird dabei bestimmt durch die sogenannte Sichtbarkeitsfunktion. Jede Person besitzt eine Sichtbarkeitsfunktion, die vom Abstand von der Position abhängt. Der Abstand wiederum wird durch die verwendete Metrik bestimmt. Die erste Wahl für die Metrik ist natürlich die Distanzmessung in Einheiten von Hypertext-Links. Aber es sind auch andere Metriken vorstellbar, wie zum Beispiel Distanzmessung durch Überlappung von Dokumenteninhalten.

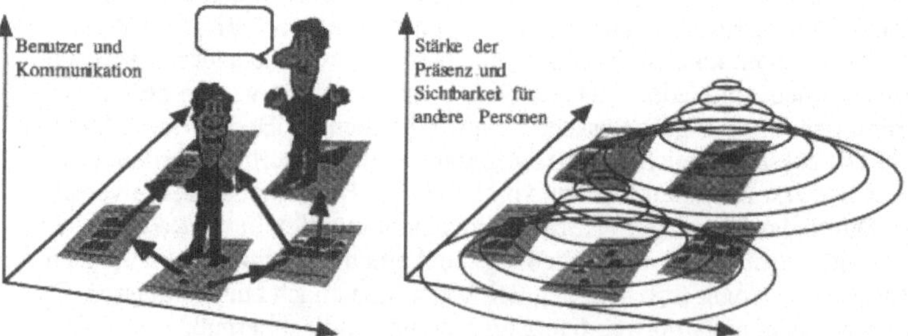

Abbildung 4:   Das Web ist die Basis der virtuellen Welt. Andere Objekte fügen neue Dimensionen hinzu. Das Web ist hier zweidimensional dargestellt, obwohl es tatsächlich ein Graph ist. Die linke Seite zeigt eine vereinfachte Darstellung von zwei Personen an verschiedenen Positionen im Web. Auf der rechten Seite sind die

entsprechenden Sichtbarkeitsfunktionen zu sehen. Die Personen können sich gegenseitig sehen und kommunizieren, wenn sich die Sichtbarkeitsfunktionen überlappen.

Kommunikationsbeziehungen können dann hergestellt werden, wenn sich die Sichtbarkeitsfunktionen überlappen. Die Kommunikation wird mit synchronen Kommunikationsdiensten, wie MBONE-tools oder Internet-Telephony, durchgeführt. Dabei sind die Eigenschaften der Kommunikation, wie Qualität, nur durch die verwendeten Programme und die Netzwerkverbindung bestimmt. So wird im Gegensatz zur realen Welt die Kommunikationsqualität (z.B. die Lautstärke) bei einer schwachen Überlappung der Sichtbarkeitsfunktionen nicht künstlich verringert.

Kommunikationsbeziehungen werden durch Kommunikationsobjekte repräsentiert. Diese Objekte fügen sich perfekt in das Modell des hier vorgestellten dynamischen Nachbarschaftsdienstes ein. Wie Personen, haben auch Kommunikationsobjekte eine Sichtbarkeitsfunktion. Die Funktion hängt dabei von den Eigenschaften der beteiligten Objekte (z.B. Sichtbarkeitsfunktionen der beteiligten Personen) ab. Ein Kommunikationsobjekt ist dann sichtbar für eine dritte Person, wenn seine Sichtbarkeitsfunktion mit derjenigen dieser Person überlappt. Zusätzlich gibt es Vorkehrungen, die Privatsphäre garantieren, falls dies von den Kommunikationspartnern gewünscht wird.

Kommunikationsobjekte dienen nur als koordinierende Instanz. Sie realisieren nicht die Kommunikation selbst. Dies bleibt externen Programmen überlassen. So ist es zum Beispiel die Aufgabe eines Kommunikationsobjekts für MBONE-tools eine Multicastadresse zu reservieren und diese an die beteiligten Kommunikationsprogramme zu verteilen. Kommunikationsobjekte für Punkt-zu-Punkt Kommunikationsprogramme, wie Internet-Phones, verteilen die Kommunikations-URLs der beteiligten Personen, z.B. RTSP URLs. Damit wird es möglich, in einem User Interface auf den Namen einer Person zu klicken, um die Kommunikationsbeziehung zu starten.

2.3 Abstraktionen

2.3.1 Objekte

Das Modell der dynamischen Nachbarschaft basiert auf einer erweiterbaren Menge von Objekttypen, bzw. Klassen im Sinne der objektorientierten Programmierung. Da das Modell der dynamischen Nachbarschaft unabhängig von der Programmiermethode ist, lehnen uns nicht zu sehr an die Terminologie der objektorientierten Programmierung an. Es werden jeweils die im Protokoll verwendeten englischen Begriffe angegeben.

Die aktuelle Implementierung umfaßt drei Objekttypen:

◆ Location (Orte)

◆ Person (Personen), und

◆ Communication (Kommunikation).

Jedes Objekt (Instanz) - einschließlich den Orten - besitzt eine Sichtbarkeitsfunktion. Diese Funktion bestimmt die Stärke der Präsenz eines Objekts im Raum, der durch die URLs aufgespannt wird.

2.3.2 Objektidentifikation

Objekte werden durch Typ und Namen identifiziert. Der Name muß global eindeutig für jeden Objekttyp sein. Er kann auf den Objekttyp zugeschnitten sein, um die Lesbarkeit des Protokolls zu erleichtern, die Adressierung externer Ressourcen zu vereinfachen oder eine effiziente Datenorganisation zu unterstützen. Orte (URLs) sind die wichtigsten externen Ressourcen. Das Konzept der URLs oder URNs wird vom Web in den Nachbarschaftsdienst übernommen. URLs werden als Namen für Ortsobjekte verwendet.

Personen- und Kommunikationsobjekte werden vom Nachbarschaftsdienst neu definiert. Für sie wird nicht wie bei Ortsobjekten auf externe Ressourcen zugegriffen. Die Namen von Personen- und Kommunikationsobjekten sind deshalb nicht durch externe Strukturen vorgegeben. Sie werden durch den Namen des Nachbarschaftsservers, der die Objekte erzeugt, global eindeutig gemacht. Der Name einer Person kann z.B. als p123@servicehost.domain:serviceport dargestellt werden. Implementierungen können aber auch andere global eindeutige Namen vergeben, wie z.B. die Email-Adresse eines Benutzers, soweit diese bekannt ist.

2.3.3 Variablen

Jedes Objekt hat Instanzvariablen, die Informationen über das Objekt speichern. Teile dieser Information werden von Benutzern bereitgestellt, z.B. in Dialogboxen eingegeben. Andere Informationen werden von Hilfskomponenten des Nachbarschaftsdienstes auf verschiedene Arten gewonnen oder zur Laufzeit erzeugt. Die veröffentlichte Spezifikation des Nachbarschaftsdienstes definiert einen Datentyp für jede Variable [6].

Die wichtigsten Variablen einer Person sind: "icon", tatsächlicher Name "realname", Position "locations" und Nachbarn "neighbors". Die Variablen "icon" und "realname" dienen dazu, die Person an der Benutzerschnittstelle darzustellen. Sie werden von den Benutzern angegeben, bzw. auf Standardwerte gesetzt, falls dies nicht geschieht. Andere Variablen kontrollieren die Berechnung des sichtbaren Bereiches und der Sichtbarkeitsfunktion. Die "neighbors"-Variable wird regelmäßig zur Laufzeit berechnet. Sie enthält die wichtige Liste der sichtbaren Personen in der Nachbarschaft.

Ortsobjekte haben Variablen, wie "icon", "links" und "persons". Das Icon dient ähnlich wie bei Personen zur Darstellung im User Interface. Es enthält z.B. ein verkleinertes Abbild einer Web-Seite (Thumbnail-Image). Die "links"-Variable enthält eine Liste von URLs, die den Hypertext-Links eines Dokuments entsprechen. Zusätzlich enthält diese Liste Linkeigenschaften wie die Distanz. Die "persons"-Variable ist das Gegenstück zur "locations"-Variable der Personen. Dies bedeutet aber nicht, daß die Information doppelt gespeichert ist. Der Inhalt der "persons"-Variable kann auch zur Laufzeit erzeugt werden. Derartige Implementierungsdetails werden von Leistungsüberlegungen bestimmt.

2.3.4 Methoden

Auf Objekte und ihre Variablen wird durch eine Menge von Methoden zugegriffen. Momentan sind folgende Methoden definiert:

- GET: Abfrage eines Variableninhalts,
- PUT: Setzen eines Variableninhalts,
- ADD und DELETE: teilweises Setzen eines Variableninhalts,
- SUBSCRIBE: Abonnieren von Meldungen über Änderungen von Variableninhalten,
- UNSUBSCRIBE: Beenden eines Abonnements,
- ASSOCIATE und DISSOCIATE: siehe 2.3.5.

Informationen zwischen Komponenten des Nachbarschaftsdienstes werden angefordert und ausgetauscht durch Kommandos. Die meisten dieser Kommandos beziehen sich auf Variableninhalte. Kommandos zur Manipulation von Variablen bestehen jeweils aus:

- Objekttyp,
- Objektnamen,
- Variablennamen,
- Methode und
- Daten (optional).

2.3.5 Assoziationen

Die Methoden ASSOCIATE und DISSOCIATE arbeiten auf Objekten. Sie werden verwendet um Objekte in Beziehung zu setzen. Das hier vorgeschlagene Nachbarschaftsmodell verwendet gerichtete, zweiwertige Assoziationen. Jede Assoziation besteht aus zwei Objekten, dem Quell- und dem Zielobjekt.

ASSOCIATE-Kommandos sind an das Zielobjekt gerichtet. Sie bestehen aus der Identifikation des Zielobjekts, der des Quellobjekts und zusätzlichen Attributen, die Details der Assoziation vermitteln. Die Objekttypen sind dabei beliebig.

Zur besseren Lesbarkeit des Protokolls wurden Parallelbezeichungen (Aliases) eingeführt, die zwischen bestimmten Kombinationen von Objekttypen anstatt des ASSOCIATE-Kommandos verwendet werden können.

Location-Location: LINK und UNLINK bedeuten Hypertext-Links zwischen Dokumenten. Dies ist ähnlich zu den Kommandos, die in Systemen zur Hyperlinkverwaltung (Distributed Hyperlink Maintenance Systems) verwendet werden, z.B. [7].

Person-Location: ENTER und LEAVE Kommandos werden beim Betreten und Verlassen von Orten (URLs, z.B. Web-Seiten) ausgetauscht. Die genauen Zeiten dieser Ereignisse werden von Hilfskomponenten des Nachbarschaftsdienstes gewonnen. Personen können gleichzeitig bei mehreren Positionen registriert sein, d.h. auf mehreren Web-Seiten gleichzeitig präsent sein.

Person-Communication: JOIN und LEAVE ordnen Personen Kommunikationsbeziehungen zu.

Person-Person: SHOW und HIDE lassen Personen selektiv sichtbar oder unsichtbar für andere werden.

Assoziationen arbeiten auch auf symbolischen Objektnamen. Momentan existieren die symbolischen Namen "_new", "_any", und "_visible". Sie bedeuten, daß ein Kommando ein neues Objekt des genannten Typs anlegen soll, von allen Objekten ausgeführt werden soll oder sich auf alle sichtbaren Objekte bezieht. Diese symbolischen Objektnamen wurden zur Optimierung eingeführt, nachdem sich gezeigt hatte, daß derartige Gruppenoperationen häufig vorkommen.

3   Realisierung

3.1   Komponenten

Der beschrieben Dienst zur Berechnung der Nachbarschaft wird vernünftigerweise als verteiltes System realisiert. Die eigentlichen Nachbarschaftsserver bilden dabei das Herz des Systems. Klienten dienen als Benutzerschnittstelle und zusätzliche Hilfskomponenten sammeln Informationen über Benutzer und das Web.

3.1.1 Server

Die Nachbarschafts-Server stellen die Benutzer-Dimension der virtuellen Welt zur Verfügung. Sie versorgen Nachbarschafts-Klienten im selben Sinne wie Web-Server die Web-Browser als Klienten bedienen. Das Web ist in Bereiche aufgeteilt, die im wesentlichen von den Web-Servern gebildet werden. Ein Nachbarschafts-Server ist verantwortlich für den Benutzerraum eines solchen Teil des Web und erzeugt den entsprechenden Teil des Benutzer-Raums. Jeder WWW-Server wird also im Prinzip durch einen Nachbarschafts-Server ergänzt, obwohl auch der Benutzer-Raum mehrerer Web-Server von einem einzigen Nachbarschafts-Server erzeugt werden kann.

Ein Nachbarschafts-Server unterhält im wesentlichen zwei Datenbanken, eine Link- und eine Benutzerdatenbank. Die Linkdatenbank spiegelt die Dokumenten- und Verbindungsstruktur des begleiteten Web-Servers wider, die noch durch Rückwärtsreferenzen aller Hypertextzeiger ergänzt wird, um symmetrische Sichtbarkeit zu gewährleisten. Links zu Objekten auf dem lokalen Server (core links) sind dabei einfach zu verwalten. Links zu Dokumenten auf anderen Servern (surface links) erzeugen hingegen Netzwerkverkehr zwischen den Nachbarschafts-Servern. Der oben beschriebene Link-Befehl (ASSOCIATE) dient zur Bekanntgabe eines Links. Er wird aber eher selten verwendet, so daß die erzeugte Netzlast neben dem HTTP-Verkehr zwischen Web-Klienten und -Servern fast verschwinden wird.

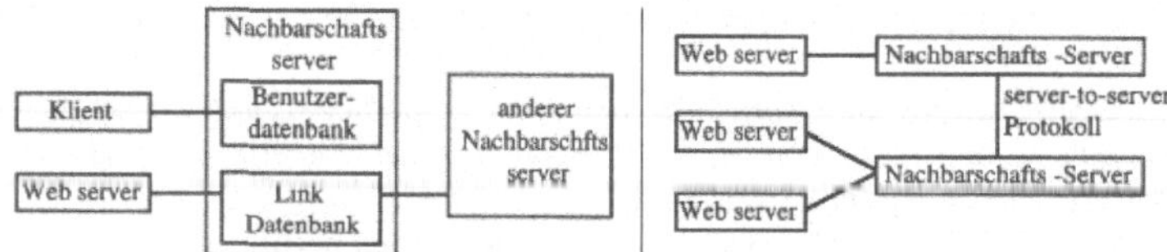

Abbildung 5: Auf der linken Seite ist eine vereinfachte Sicht eines Nachbarschafts-Servers dargestellt. Er kommuniziert mit Klienten, Web-Servern und anderen Nachbarschafts-Servern. Der rechte Teil zeigt eine Situation, in der ein N-Server für mehrere Web-Server verantwortlich ist.

Die Hauptaufgabe eines Nachbarschafts-Servers ist die Berechnung von sichtbaren Objekten. Sie geschieht auf der Grundlage der Link-Struktur, der Benutzervorgaben und verschiedener Systemparameter. Benutzervorgaben und Server-spezifische Einstellungen beeinflussen die Berechnung der Sichtbarkeitsfunktion und somit der Menge der sichtbaren Objekte. Gegenwärtig benutzen wir eine in Raum und Zeit quaderförmige Funktion: die personenbezogene Variable Link-Distanz bestimmt die Länge des Quaders. Der Wert 2 bedeutet zum Beispiel, daß alle Personen, die maximal zwei Hypertext-Referenzen entfernt sind, sichtbar sind (siehe Abbildung 6).

Ein zweiter Parameter steuert die Tiefe der Sichtbarkeit in der Zeit. Obwohl die Zeit nicht in Abbildung 4 gezeigt wurde, hat sie sich doch als sehr sinnvoll für die Benutzbarkeit des Systemes erwiesen. Die zeitliche Dimension der Sichtbarkeitsfunktion glättet nämlich die Bewegung der Benutzer zwischen Seiten, Benutzer bleiben noch für eine Weile sichtbar, obwohl sie die Seite schon verlassen haben.

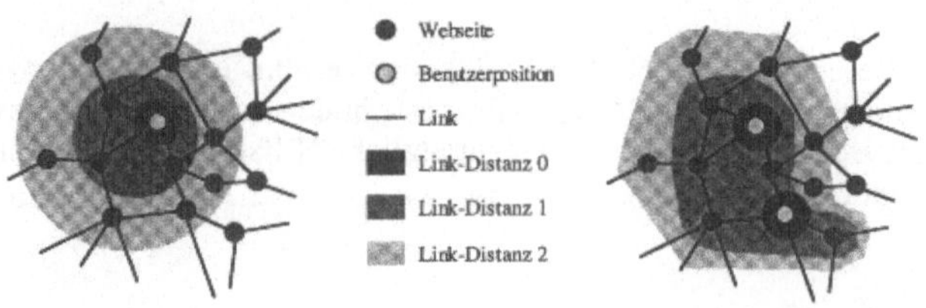

*Abbildung 6:   Seiten im Sichtbarkeitsgebiet hängen von der Linkdistanz ab. Im rechten Teil
kann man sehen, daß die Form dieser Zone durchaus komplex werden kann, da
Benutzer mehrere Seiten gleichzeitig besuchen können.*

Das Resultat der Berechnung, die Liste der sichtbaren Personen, Seiten und der Kommunikationsbeziehungen wird für die Nachbarschafts-Klienten, beispielsweise die Komponenten der Benutzerschnittstelle (das Userinterface), bereitgehalten.

3.1.2   Klienten

Nachbarschafts-Klienten präsentieren dem Benutzer eine Sicht auf den Benutzerraum. Sie benutzen das oben eingeführte Protokoll um Information über den Aufenthaltsort der Benutzer, die Umgebung, andere Benutzer und deren Eigenschaften abzufragen.

3.1.2.1   Benutzerschnittstelle

Die einfache Benutzerschnittstelle hat als Testwerkzeug für die Protokollmaschine und den Nachbarschafts-Server begonnen. Im Laufe der Entwicklung hat sie sich aber zum 'fettfreien' User Interface gemausert, das sich durch kurze Ladezeiten über das Internet und einfachste Bedienung auszeichnet. Es zeigt die Nachbarn als Liste mit Bildern, Namen und Position (URL). Dieses einfache Interface erlaubt auch die Änderung der Instanzvariablen (siehe 2.3.3) durch den Besitzer eines Objektes im Nachbarschafts-Server.

*Abbildung 7:   Die einfache Benutzerschnittstelle: Nachbarn werden als kleine Bilder (Icons)
gezeigt, die auch zum Start von Kommunikationsverbindungen angeklickt werden können. Das kleine Chat-Fenster rechts oben soll die Benutzer zur Teilnahme ermuntern.*

3.1.2.2   Erweiterte Benutzerschnittstelle

Die Virtual Reality Modeling Language (VRML) wurde in der Version 2.0 durch das sogenannte external authoring interface (EAI) erweitert, so daß dynamische virtuelle Welten dargestellt werden können. Anders als die einfache Benutzerschnittstelle zeigt unser erweitertes User Interface nicht nur eine Liste der Benutzer. Es modelliert vielmehr einen Ausschnitt des WWW und des Benutzerraumes dreidimensional, ähnlich wie das in [8], "A Graphical Hypertext Navigation Tool" geschieht. Es verwendet aber die VRML-Maschine, um die dreidimensionalen Szenen auf dem Computer des Benutzers anzuzeigen. Die erweiterte Benutzerschnittstelle kommuniziert dabei mit dem Modell über dynamische VRML-Objekte und Java-Skripts. Um die Umgebung besser zu symbolisieren, werden verkleinerte Abbilder der WWW-Seiten, sogenannte Thumbnails, von einem neuartigen Thumbnailserver erzeugt. Diese Repräsentanten der 'location objects' werden durch Linien verknüpft, die Links symbolisieren. Wieder dienen Icons und Namen als Platzhalter für Personen, sie werden um die Seiten, die sie gerade besuchen, gruppiert. Textkommunikation zwischen Benutzern (chat) erscheint in dieser Schnittstelle als Textspur, die vom Vordergrund der 3-D-Szene in der Hintergrund fließt. Wieder kann man mit der Maus auf die Bilder der Personen drücken, um Audio- und Video-Kommunikation einzuleiten.

3.2 Protokoll

Die Gesamtheit der Protokolldaten, Befehle und Parameter werden in HTTP gekapselt. Der wichtigste Vorteil dieser Technik ist die Kompatibilität mit anderen WWW- und Internet-Komponenten, wie zum Beispiel Proxies und Firewalls.

Das Nachbarschafts-System, Server und Klienten, basiert auf Daten, die sich ständig ändern. Um die Nachbarschaft korrekt zu berechnen und die Sichtbarkeit anzuzeigen, müssen Änderungen in der Datenbank oder von Parametern anderen Komponenten zügig gemeldet werden. Hier unterscheidet sich das dynamische Wesen des Nachbarschaftsdienstes ganz wesentlich von der statischen Natur des WWW. Während der Implementierungsarbeiten wurde klar, daß das Protokoll für den Transport solcher Information ein Dienstelement zum dynamischen Update benötigt, das im klassischen HTTP-Request/Response Verfahren nicht vorgesehen ist.

Wir haben uns deshalb entschieden, einen asynchronen Mechanismus in das HTTP Request/Response Verfahren einzubetten. Dabei wird die Methode SELECT verwendet, um Änderungsbenachrichtigungen für ein bestimmtes Objekt zu abonnieren. SELECT erzeugt einen wartenden Request beim Empfänger, der entweder nach einem sehr langen Timeout erfolglos oder vorher mit einer Änderungsbenachrichtigung zurückkehrt. Der Abonnent wird den neuen Wert anschließend mit einer konventionellen Request/Response Transaktion abholen, so daß kein unnötiger Verkehr durch regelmäßig wiederkehrende Abfragen (Polling) entsteht.

4 Ergebnisse

Der in diesem Bericht beschriebene Dienst wurde von uns in den Jahren 1996 und 1997 implementiert. Der voll funktionsfähige Prototyp wurde erstmals im Oktober 1997 auf mehreren WWW-Servern installiert und funktioniert seitdem zur vollen Zufriedenheit.

Diese Server berechnen unter anderem die Nachbarschaft für zwei besondere WWW-Server: eine Beispiel-WWW-site, den virtuellen Automobilhandel. Besucher können Fahrzeuge besichtigen, technische Daten abfragen und mit einem Verkäufer sprechen.

die interaktive Modellbahn (http://rr-vs.informatik.uni-ulm.de/) hat mehr als 1000 Besucher pro Tag. Diese Besucher kommen, um mit der Modellbahn zu spielen und um im Echtzeit-Video zu sehen, wie die Lokomotiven auf Befehl ihre Kreise ziehen. Seit Oktober können sie nun auch andere Spieler sehen und mit ihnen sprechen. Durch die große Anzahl Besucher ist unser Server auch einem Härtetest unterzogen worden.

Beide Installationen haben gezeigt, daß die Benutzer überrascht und erfreut sind, andere Benutzer auf der selben Seite zu sehen. Da die meisten Benutzer (noch) keine Multimediaausrüstung und die nötige Konferenz-Software haben, wird von dem eingebauten chat-Werkzeug eifrig Gebrauch gemacht. Problematisch ist allerdings die etwas umständliche Bedienung des fernschreibartigen Chat-Dienstes, die die Verweildauer in der Konferenz doch verkürzt. Obwohl sich manche Benutzer sehr bemühen, die Unterhaltung in Gang zu halten, sind andere einfach zu bequem, die Tastatur zu benutzen.

Die Implementierungen der zentralen Komponenten, also des Nachbarschafts-Servers und der zugehörigen Programme zum Sammeln von Informationen, sind zuverlässig und auch schnell genug für ihre Aufgaben. Unsere anfängliche Sorge, daß einfache Graph-Traversierungsalgorithmen nicht leistungsfähig genug sind, hat sich als unbegründet erwiesen. Falls die Bearbeitung der Graphen in extrem viel besuchten, großen Servern in Zukunft ein Problem werden sollte, können besonders effiziente Verfahren, wie sie in der OR-Forschung entwickelt worden sind, Abhilfe schaffen.

Die Implementierung der Dienstelemente, die in die Klienten-Computer geladen werden, also die Benutzerschnittstelle, hat uns hingegen deutlich mehr Schwierigkeiten gemacht. Als Applets sind sie zeitgemäß in Java programmiert. Obwohl sich Java als Programmiersprache bewährt hat, hat der Test der Applets sich als äußerst zeitraubend erwiesen. Jede Komponente muß auf jedem Browser auf jedem unterstützten Betriebssystem erneut getestet und meist auch umprogrammiert werden.

5 Aktuelle und zukünftige Arbeiten

Das dynamische Nachbarschafts-System ist implementiert und wird bereits genutzt. Wie immer gibt es viele Möglichkeiten, Dienst und System zu verbessern. Wir arbeiten laufend an der Wartung und planen Erweiterungen, die die Benutzung erleichtern sollen oder Forschungscharakter haben.

5.1 Erweiterte Nachbarschaftsmodelle

Die Benutzer sollten die Möglichkeit haben, die Berechnung ihrer persönlichen Sichtbarkeits-Funktion zu steuern. Das gilt besonders in Servern, in denen sich Hunderte von Benutzern gleichzeitig aufhalten. In anderen Fällen wollen die Benutzer nur für eine bestimmte Gruppe sichtbar sein, die dafür aber auch weiter entfernt sein können. Deshalb wollen wir die Berechnung der Nachbarschaft durch benutzerspezifische Programme - Meetlets - erweitern.

Theoretisch genügen benutzerspezifische Programme, um die Sichtbarkeitsfunktion zu berechnen. In der Praxis ist das allerdings undurchführbar, da diese Methode wegen ihrer Komplexität ($O(n2)$) nicht skalierbar ist auf Größenordnungen von Tausenden oder mehr Benutzern . Deshalb benötigt der Dienst eine relativ kleine Startmenge, die auf konventionelle Weise berechnet wird. Die Meetlets dieser Menge erzeugen dann eine Untermenge, die sichtbar wird. Wir haben mit der Verwirklichung eines solchen Ansatzes auf der Basis unseres existierenden Systems begonnen. Die Meetlets entscheiden, ob ein Objekt durch den Benutzer gesehen werden soll und ob Objekte des Benutzers, zu dem das Meetlet gehört, für andere sichtbar sein sollen.

5.2 Benutzbarkeit und Installierbarkeit

Einfache Benutzbarkeit entsteht aus intuitiver Visualisierung der Nachbarschaft und direkter Manipulation der Kommunikationsbeziehungen. Diese Eigenschaften werden in der fortgeschrittenen Benutzerschnittstelle verwirklicht. Der Preis für diese Weiterentwicklung ist jedoch hoch: der Programmcode wird kompliziert und vor allem groß. Im Falle eines zur Laufzeit über das Internet geladenen Applets, das auf der Zielmaschine interpretiert werden muß, entsteht beim Benutzer beträchtlicher Unmut über die langen Wartezeiten. Obwohl wir gegenwärtig das Verhältnis Benutzbarkeit-Codegröße optimieren, verfolgen wir gleichzeitig einen anderen Ansatz, nämlich die Entwicklung eines konventionelle Programmes, das auf dem Klienten-Rechner permanent installiert wird.

Für Betreiber bedeutet Benutzbarkeit einfache Installation des Dienstes auf existierenden WWW-Sites, ohne daß der Inhalt neu geschrieben oder überarbeitet werden muß. Unser System kann bisher als zusätzlicher Dienst unter WindowsNT mit dem IIS oder unter UNIX mit dem CERN-Server mit minimalem Aufwand installiert und betrieben werden. Gegenwärtig erstellen wir Filter für den Netscape Enterprise Server und ein Apache Modul. Zur Wartung der Komponenten ist bereits ein WWW-basiertes Management Interface eingebaut.

6 Zusammenfassung

In den vorangegangenen Kapiteln haben wir Konzept, Implementierung und erste Ergebnisse unseres Systems zur dynamischen Nachbarschaft im World Wide Web vorgestellt. Das System ist implementiert und befindet sich seit einigen Monaten im Einsatz auf unseren WWW-

Servern. Ungefähr 1000 Besucher werden pro Tag bearbeitet. Diesen Benutzern wird ihre Nachbarschaft während des Besuches graphisch präsentiert. Sie können andere sehen und deren Bewegung von Seite zu Seite beobachten. Außerdem erhalten sie Gelegenheit zur synchronen Kommunikation untereinander - ein einfacher Druck auf die Maustaste startet ein Internet-basiertes Videotelefon.

Wir warten und verbessern das System ständig. Außerdem erweitern wir den Dienst durch neue, verbesserte Mechanismen wie die Meetlets. Der Erfolg des Dienstes hängt jedoch vor allem von seiner Installation auf möglichst vielen WWW-Servern ab. Jeder Leser ist deshalb aufgefordert, die kostenlose Software von unserer WWW-site http://www.cobrow.com zu laden und auf seinem eigenen WWW-Server zu installieren. Selbstverständlich werden wir uns im Rahmen des "Universitätsmöglichen" bemühen, die Programme zu warten und bei der Installation zu helfen.

Die beschriebene Arbeit wurde im Rahmen des EU Telematics Projektes Cobrow (RE1003) durchgeführt. Wir möchten unseren Projektpartnern danken, die an anderen Komponenten des CoBrow-Dienstes, den synchronen Internet-Konferenzwerkzeugen und den Integrationsmechanismen für CSCW-Programmen, arbeiten.

7 Literaturverzeichnis

[1] Firefly Network, Inc. Personalize your Network. Cambridge, MA, 1997, Software online at http:/www.firefly.net/.

[2] Mirabilis Ltd, ICQ - World's Largest Internet Online Communication Network, Tel Aviv, Israel, 1997, Software online at http://www.mirabilis.com/.

[3] WerWeissWas, a search engine for experts, Germany, 1997, Software online at http://www.wer-weiss-was.de/.

[4] URN Syntax. R. Moats. Internet RFC 2141, proposed standard, May 1997, ftp://ds.internic.net/rfc/rfc2141.txt.

[5] R. Fielding: Maintaining Distributed Hypertext Infostructures: Welcome to MOMspider's Web, Computer Networks and ISDN systems, Volume 27, issue 2, 1994.

[6] EU Telematics Project CoBrow: "System Specification", Project Deliverable 4.3, Feb. 1997

[7] James E. Pitkow, R. Kipp Jones: Supporting the Web: A Distributed Hyperlink Database System, Fifth International World Wide Web Conference, Paris, France, May 6-10, 1996.

[8] P. Domel: WebMap - A Graphical Hypertext Navigation Tool. Proceedings of the Second International World Wide Web Conference. Chicago: 1994. http://www.tm.informatik.uni-frankfurt.de/Publications/Doemel/WWWFall94/www-fall94.html>

Adressen der Autoren

Dr. Konrad Froitzheim
Abteilung Verteilte Systeme
Universität Ulm
89069 Ulm
Email: frz@informatik.uni-ulm.de

Klaus H. Wolf
Abteilung Verteilte Systeme
Universität Ulm
89069 Ulm
Email: wolf@informatik.uni-ulm.de

Interaktion durch die Verbindung von Telefonie-Anwendungen mit dem Web

Christoph Hochstätter

Microsoft

Eine schriftliche Zusammenfassung des Beitrages konnte nicht rechtzeitig fertiggestellt werden.

Unterstützung kooperativer Software-Wiederverwendung durch Internet-Technologien *

Anita Behle

Lehrstuhl für Informatik III, RWTH Aachen

Zusammenfassung

In anderen Industriebereichen ist er bereits Alltag, im Bereich der Software-Entwicklung setzt er sich nur langsam durch: Der Einsatz von fertigen, wiederverwendbaren Bausteinen bei der Entwicklung neuer Produkte. In diesem Artikel wird ein Ansatz zur kooperativen Wiederverwendung von Software-Komponenten im Rahmen eines Verbundes von Software-Herstellern vorgestellt. Es handelt sich dabei um ein kooperatives Software-Komponenten-Informationssystem im World Wide Web, welches einerseits über Software-Bausteine der beteiligten Unternehmen, andererseits über öffentlich verfügbare Bausteine informiert und einen Austausch von Komponenten und Erfahrungen ermöglicht.

Vorgestellt werden die Konzepte und die Realisierung des REGINA Komponenten-Informationssystems, welches in Zusammenarbeit mit mehreren Industriepartnern entwickelt wird und bereits im Internet verfügbar ist.

1 Einleitung

Obwohl die Vorteile der Entwicklung von Software unter Wiederverwendung von bereits vorhandenen Software-Bausteinen allgemein bekannt sind, ist die Einführung von Mechanismen zur Unterstützung der Wiederverwendung in Unternehmen oft problematisch.

Das Hauptproblem liegt darin, daß Software, die für ein System entwickelt wurde, im allgemeinen nicht ohne Probleme in ein neues System eingebaut werden kann, da sie eine „Spezialanfertigung" ist und bei ihrer Erstellung ein späterer Einbau in ein anderes System nicht berücksichtigt wurde.

Man muß also die Software-Entwicklung *mit* Wiederverwendung (Development *with* Reuse) von der Software-Entwicklung *für* Wiederverwendung (Development *for* Reuse) unterscheiden [10]. Die Entwicklung von leicht wiederverwendbarer Software ist zeit- und kostenintensiv. Biggerstaff hat dazu die „Rules of three" aufgestellt [13]: Einerseits muß Software dreimal entwickelt werden, bevor sie wirklich wiederverwendbar entwickelt werden kann, und andererseits muß man Software-Bausteine mindestens dreimal wiederverwenden, bevor man von der Wiederverwendung profitiert.

Bisher haben daher hauptsächlich große Software-Hersteller mit eigenen Forschungseinrichtungen Wiederverwendungsprojekte durchgeführt [15]. Die Gründe sind die längere Dauer

---

* Das Projekt wird gefördert mit Mitteln des Ministeriums für Wirtschaft und Mittelstand, Technologie und Verkehr (MWMTV) sowie des Ministeriums für Wissenschaft und Forschung (MWF) des Landes Nordrhein-Westfalen.

und die damit verbundenen höheren Kosten für die erste Erstellung eines wiederverwendbaren Bausteins. Berichte über Produktivitätssteigerungen und Qualitätsverbesserungen [11] ermutigen jedoch dazu, auch bei kleineren Betrieben langfristig Software-Entwicklung mit und für Wiederverwendung einzuführen.

Kleine oder mittelständische Software-Häuser führen hauptsächlich Projektgeschäfte durch. Die Zeiträume für Projekte sind im allgemeinen fest abgesteckt und die gesamte Software-Entwicklung erfolgt streng projektbezogen. So ist es normalerweise nicht möglich, allgemeine Bausteine zu entwickeln, die in mehreren Projekten genutzt werden können. Auch bei eigenen Entwicklungen ist selten vorauszusehen, daß Software in Zukunft noch einmal genutzt werden könnte und sich so die Entwicklung eines allgemeinen Bausteins wirtschaftlich rentieren würde. Aus diesen Gründen liegen häufig nur wenige Software-Bausteine in einer Qualität vor, die eine Wiederverwendung ermöglichen würde.

Welche Möglichkeiten bestehen also für kleine und mittelständische Betriebe, die keine oder nur sehr kleine Software-Baustein-Bibliotheken besitzen? Naheliegend ist die Nutzung von *fremden* Software-Bausteinen, deren Qualität einen unkomplizierten Einbau in ein neues System zuläßt und deren Kosten unter denen der Eigenentwicklung liegen. Hinzu kommt bei der Nutzung fremder Bausteine, daß die Zeit für die Entwicklung gespart werden kann und so kürzere Entwicklungszyklen entstehen.

Die Förderung dieses Ansatzes ist das Ziel des RSB-Projektes (REGINA Software-Bibliothek), welches vom Lehrstuhl für Informatik III wissenschaftlich begleitet wird und in Kooperation mit neun Mitgliedern von REGINA durchgeführt wird. REGINA (Regionaler Industrie-Club Informatik Aachen e.V.) ist ein Verein, der 1991 gegründet wurde und ein Forum für die regionale Kooperation im Informatikbereich in der Wirtschaftsregion Aachen darstellt. Im Rahmen dieses Projektes wird seit April 1995 ein Informationssystem entwickelt, das im World Wide Web (WWW) angesiedelt ist und umfassend über wiederverwendbare Software-Bausteine sog. Komponenten informiert, die bei kooperierenden Software-Herstellern oder auf dem Software-Markt erhältlich sind und bei der Entwicklung neuer Software eingesetzt werden können. Bei den beschriebenen Software-Bausteinen handelt es sich sowohl um binäre Komponenten (JavaBeans, ActiveX, VBX, DLL, ...) mit definierter Schnittstelle als auch um Klassenbibliotheken. Beide Arten können als Baustein bei der Entwicklung eines neuen Software-Systems eingesetzt werden.

Nachfolgend werden zunächst die Anforderungen an das Komponenten-Informationssystem (KIS) vorgestellt, welche in Zusammenarbeit mit den Industrie-Partnern festgelegt wurden. Anschließend werden die Konzepte, die Funktionalität des Systems sowie die Realisierung auf der Basis verschiedener Internet-Technologien präsentiert. Dabei wird besonders auf die verschiedenen Interaktionsmöglichkeiten und die dabei eingesetzten Internet-Technologien eingegangen. Zum Abschluß werden kurz einige verwandte Arbeiten vorgestellt.

2 Anforderungen

Die Anforderungen an das Software Komponenten-Informationssystem wurden zusammen mit den Industriepartnern festgelegt und betreffen hauptsächlich die Informationen, die durch das System bereitzustellen sind (Inhalt), sowie die prinzipiellen Registrierungs- und Zugriffsmöglichkeiten, die Benutzern und Administratoren zur Verfügung gestellt werden müssen (Benutzerschnittstelle).

2.1 Benutzerschnittstelle

Zunächst sollte das System für alle Projektteilnehmer erreichbar sein, so daß sehr schnell klar war, daß das System im Internet bzw. im World Wide Web anzusiedeln ist und damit lediglich ein Standard Web-Browser für den Zugriff benötigt wird. Weiterhin sollten die Informationen nur für registrierte Benutzer zugänglich sein. Zur Handhabung verschiedener Zugangsrechte und Sichten, beispielsweise externe Benutzer (zugelassene Gäste) oder Projektmitglieder, wird eine Verwaltung für verschiedene Benutzergruppen benötigt.

Die Benutzerschnittstelle sowohl für die Registrierung und Pflege von Benutzer- und Komponentendaten als auch zur Suche nach Informationen sollte benutzerfreundlich, intuitiv und einheitlich sein. Zur Suche nach Informationen sind außerdem verschiedene Möglichkeiten anzubieten, welche je nach Wissen über die gesuchte Komponente genutzt werden können.

2.2 Inhalt

Zweck des Komponenten-Informationssystems ist der Austausch von konkreten bei den einzelnen Projektpartnern vorhandenen Software-Komponenten sowie der Austausch von Informationen und Erfahrungen über Bausteine, die von Dritten angeboten werden, also frei erhältliche oder kommerziell verfügbare Produkte. So kann sich ein Software-Entwickler an einer zentralen Stelle umfassend über sämtliche ihm zur Verfügung stehenden Software-Komponenten informieren.

Im Gegensatz zu anderen Wiederverwendungsansätzen werden in diesem System Informationen zu Komponenten bereitgestellt, nicht die Komponenten selbst. Bei den Daten zur Beschreibung einer Komponente handelt es sich um eine große Menge unterschiedlichster Informationen (Name, Funktionalität, unterstützte Plattformen und Compiler, Bezugsquellen, Anwendungsbereiche, Umfang der Dokumentation, ...). Dabei können Komponenten beschrieben werden, die in unterschiedlichen Programmiersprachen implementiert wurden und sich auf verschiedenen Granularitätsstufen befinden (einzelne Klassen, Klassenbibliotheken, binäre Komponenten, aber auch ganze Frameworks).

Wesentlich ist, daß das System so konzipiert wird, daß Informationen auf einfache Art geändert, gelöscht oder umstrukturiert werden können, da gerade im Bereich der Software-Entwicklung ständig neue Anwendungsgebiete und Technologien hinzukommen.

Die Beschreibung einer Komponente muß komplett auf einer Seite in einer übersichtlichen, tabellarischen Form angezeigt werden, welche alle Informationen stets mit dem gleichen Vokabular in der gleichen Reihenfolge enthält. Nur so kann der Entwickler einen schnellen Überblick erhalten und verschiedene Komponenten auf einfache Art vergleichen.

3 Konzepte und Realisierung

Im folgenden werden kurz die wichtigsten Konzepte des Komponenten-Informationssystems zusammen mit ihrer Realisierung vorgestellt. Dies sind der Klassifikationsansatz, die dynamisch generierte Benutzerschnittstelle und die Integration in das World Wide Web. Die Integration des Komponenten-Informationssystem in eine globale Software-Wiederverwendungsumgebung wurde in [3] vorgestellt und wird hier nicht weiter behandelt.

Da das KIS im Internet verfügbar ist (*http://www.findcomponents.com*), werden bis auf eine Ausnahme keine Screendumps vorgestellt.

Um die Übertragbarkeit der Konzepte und Realisierung des Komponenten-Informationssystems auf Anwendungen bei den Projektpartnern zu gewährleisten, wurde gefordert, kommerzielle und portable Software zur Realisierung einzusetzen. Wir haben uns daher für eine relationale Datenbank und einen WebServer von Oracle für die Plattform Windows NT entschieden. Der Oracle7 WorkgroupServer stellt das RDBMS dar, in welchem alle Daten gespeichert werden. Der Oracle WebServer stellt die Verbindung zwischen Internet und der Datenbank dar. Zum Zugriff auf die Datenbank wurden in Zusammenarbeit mit einigen Projektpartnern PL/SQL-Routinen mit eingebettetem SQL entwickelt, welche über den sog. Oracle Web Agent ähnlich wie CGI-Skripte gestartet werden können.

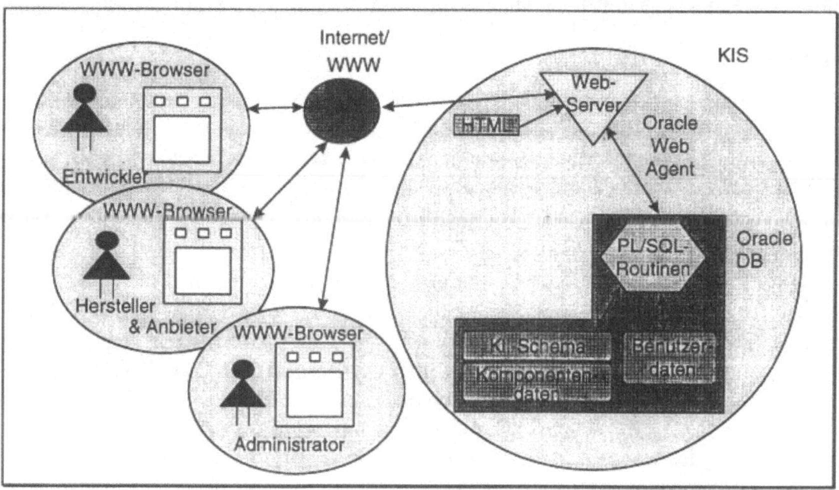

Abbildung 1: Szenario der Client/Server Architektur des KIS im Internet

Abbildung 1 zeigt eine Darstellung der Client/Server-Architektur des Komponenten-Informationssystems im Internet. Auf der linken Seite sind die drei potentiellen Benutzergruppen zu sehen:
1. der eigentliche Nutzer des Systems, also ein Entwickler in einem Software-Unternehmen, der Software sucht,
2. Hersteller und Anbieter von kommerziellen oder frei erhältlichen Software-Komponenten, die ihre Produkte registrieren oder die eingetragenen Daten pflegen möchten,
3. der Administrator, der neben dem Klassifikationsschema und den Benutzerdaten alle anderen Daten ebenfalls pflegen kann.

Auf der rechten Seite ist das Komponenten-Informationssystem mit seinen Bestandteilen zu sehen. Über den Web-Server wird das System an das Internet angeschlossen. Die Zugriffsroutinen und sämtliche Daten werden in der Oracle Datenbank abgespeichert. Diese Bestandteile werden im folgenden näher erläutert. Neben dem Zugang über den Web-Server und HTTP gibt eine weitere Zugriffsmöglichkeit über eine Java Client/Server Anwendung, welche im Anschluß vorgestellt wird.

3.1 Das Klassifikationsschema

Abgesehen von einer benutzerfreundlichen Oberfläche, die im nachfolgenden Abschnitt vorgestellt wird, ist die Bereitstellung von möglichst detaillierten und vergleichbaren Informationen Voraussetzung für ein attraktives Informationssystem. Die Darstellung eines reinen Beschreibungstextes ist dazu nicht geeignet. Statt dessen wird der Ansatz der Software-Klassifikation als Grundlage sowohl für die Datenspeicherung als auch für die Erstellung der Produktbeschreibungen verwendet. Wir nutzen dazu eine Kombination der facettierten Klassifikation [12] und der in [7] vorgeschlagenen Klassifikationshierarchie. Wir kombinieren so die Vorteile der Klassifikation auf der Basis von Facetten und vordefiniertem Vokabular mit einer hierarchischen Struktur, welche im Gegensatz zu einer flachen Struktur Klarheit in die Zusammenhänge zwischen den Facetten und Merkmalen bringen kann. Außerdem berücksichtigt das Konzept neben den Facetten und vordefinierten Termen die Möglichkeit, kurze Texte zur Spezifikation spezieller Aspekte anzugeben (z.B. Lizenzbedingungen oder WWW-Links).

Da aufgrund der starken Dynamik im Software-Bereich zu erwarten ist, daß das Klassifikationsschema häufig erweitert, geändert oder umstrukturiert werden muß, ohne daß bestehende Datenbestände dadurch gefährdet werden dürfen, wurde der Ansatz eines Meta-Schemas gewählt. Dieses Meta-Schema beschreibt das Klassifikationsschema und die Informationen zu Software-Komponenten, die durch das Klassifikationsschema beschrieben werden können.

Das Meta-Schema bildet die Grundlage des (relationalen) Datenbankschemas. Es wurde in [4] ausführlich beschrieben. So können sowohl das Klassifikationsschema als auch die korrespondierenden Komponentendaten in der Datenbank gespeichert und einfach verändert werden. Unsere Untersuchungen und praktischen Tests haben dabei gezeigt, daß nur bei wenigen Schemaänderungen eine Interaktion mit dem Administrator des Systems notwendig ist.

3.2 Die Benutzerschnittstelle

Wie bereits erwähnt, ändert sich das Klassifikationsschema häufig. Da das Klassifikationsschema jedoch die Grundlage der Benutzerschnittstelle darstellt, muß die Oberfläche jedesmal angepaßt werden, wenn das Klassifikationsschema verändert wurde. Um den Aufwand einer manuellen Anpassung zu umgehen, wird die Oberfläche zur Laufzeit auf der Basis des Klassifikationsschemas und des Datenbestandes dynamisch generiert. So sind Änderungen des Klassifikationsschemas schon beim nächsten Zugriff auf die Benutzeroberfläche sichtbar.

Für die bereits vorgestellten Benutzergruppen werden nun verschiedene Interaktionsmöglichkeiten mit dem System angeboten. Alle Dialoge der Benutzeroberfläche, die als HTML-Seiten mit HTML-Formularen angeboten werden, besitzen dabei den gleichen Aufbau, welcher durch allgemeine Layouteinstellungen vorgegeben wird. Die Layouteinstellungen sind an einer zentralen Stelle des Systems definiert, so daß Änderungen des Layouts der gesamten Oberfläche nur an einer Stelle vorzunehmen sind.

Entwickler können je nach ihrem Wissensstand auf verschiedene Arten nach Komponenten suchen. Dies wird einerseits durch eine Stichwortsuche ermöglicht, bei welcher zusätzlich der

Suchbereich auf Plattformen und Anwendungsgebiete eingeschränkt werden kann. Diesen Zugang zeigt Abbildung 2. Andererseits kann die gesamte Datenbank durch Navigation erforscht werden. Es werden dazu verschiedene Einstiegspunkte angeboten, wie beispielsweise die Anzeige nach Namen, Plattformen, Herstellern oder Anbietern, Komponentenarten usw. Die logische Verknüpfung von mehreren Suchbegriffen und der Einsatz von Information Retrieval (IR) Techniken zur besseren Suche nach Begriffen sind vorgesehen, jedoch zur Zeit noch nicht implementiert. Die Eingabe von Erfahrungen und Referenzen zum Erfahrungsaustausch über Komponenten befindet sich zur Zeit in Bearbeitung.

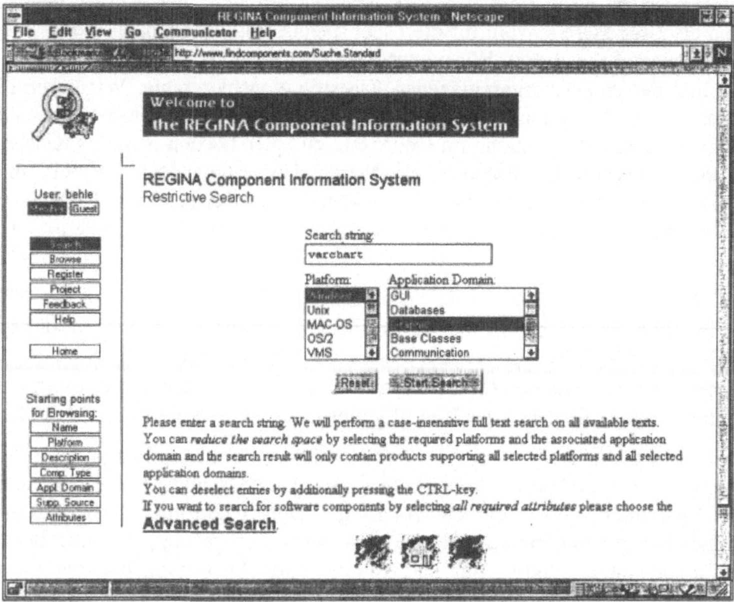

Abbildung 2: Screendump eines HTML-Formulars zur Suche

Mitglieder des Konsortiums und auch andere Hersteller und Anbieter können die von ihnen angebotenen Software-Komponenten registrieren und pflegen. Dazu erhalten die Unternehmen nach der ersten Registrierung ein Paßwort, welches ihnen erlaubt, die Daten der von ihnen betreuten Komponenten zu ändern. Sollte in dem Klassifikationsschema ein Eintrag (beispielsweise ein charakteristisches Merkmal eines Anwendungsgebietes) fehlen, kann der Eintrag in einer weiteren Seite zur Aufnahme in das Klassifikationsschema vorgeschlagen werden. Der Administrator hat neben der Pflege der Benutzer-, Hersteller- und Komponentendaten außerdem die Aufgabe, das Klassifikationsschema zu warten.

Die Oberfläche selbst besteht zum einen aus HTML-Formularen, zum anderen aus einer Java Client/Server-Applikation; beide Ansätze werden nun kurz vorgestellt.

3.2.1 HTML

Betrachtet man die Vorgehensweise des Informationsaustauschs im World Wide Web auf der Basis des Hypertext Transfer Protocols (HTTP) und der Hypertext Markup Language

(HTML), so stellt man fest, daß Anfragen mit Hilfe von HTML-Formularen an WebServer gestellt werden. Als Ergebnis auf eine Anfrage muß stets wieder eine HTML-Seite geliefert werden. Verwendet man also HTML zum Zugriff auf einen Datenbestand, müssen die Ergebnisseiten in jedem Fall dynamisch generiert werden. Auch die anderen Teile der Benutzeroberfläche, beispielsweise die HTML-Formulare selbst, werden auf diese Art erstellt.

Abbildung 3 zeigt das Prinzip der dynamischen Generierung. Die aktuellen Daten bezüglich Suchtext oder die Navigationsposition, Datenbestand, Klassifikationsschema, Layouteinstellungen und auch die Zugriffrechte des Benutzers werden bei der Generierung jeder HTML-Seite berücksichtigt.

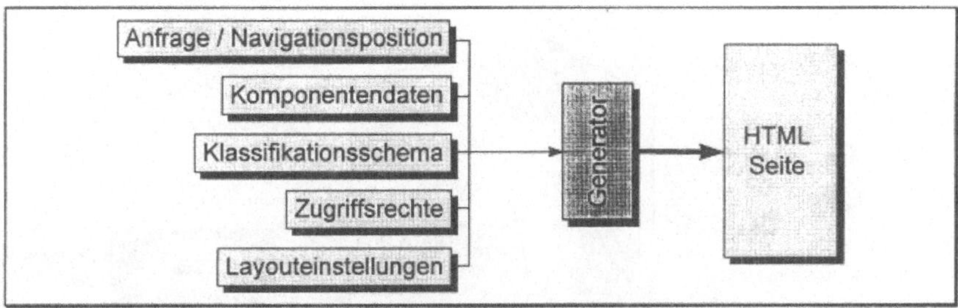

Abbildung 3: Dynamische Generierung der HTML-Seiten

Protokollelemente des HTTP erlauben es, den Benutzernamen zu ermitteln, so daß jede Seite individuell für jeden Benutzer mit einer Sicht entsprechend seinen Zugriffrechten erzeugt wird. Ein weiterer Schritt, welchen wir konzeptionell vorgesehen, jedoch noch nicht realisiert haben, ist die Analyse und Berücksichtigung von Präferenzen des Benutzers [6].

3.2.2 Java

Obwohl die Nutzung von HTML-Formularen für Anfragen im World Wide Web weit verbreitet ist, hat diese Vorgehensweise auch einige Nachteile. Zum einen sind sie nur schlecht zur Darstellung komplexer Sachverhalte geeignet. Der Aufbau großer Tabellen, welche in der Regel benutzt werden, um umfangreiche Daten übersichtlich darzustellen, ist zeitaufwendig. Außerdem kann eine komplexe Eingabe nur schlecht in mehrere kleinere Dialoge aufgesplittet werden, da der Web-Server generell zustandslos ist und so keine Aktionen über mehrere Seiten verwalten kann. Man muß sich dann umständlicher Methoden (z.B. sog. Hidden-Werte) bedienen oder eine große und unübersichtliche HTML-Seite zur Eingabe der Daten anbieten. Bei diesem Problem liefern auch VBScript und JavaScript keine Lösung, welche in HTML eingebettet werden können, um Plausibilitätskontrollen durchzuführen und festzustellen, ob ein HTML-Formular korrekt ausgefüllt ist.

Da wir aber trotzdem nicht auf das Internet und den WWW-Browser als grundlegende Software verzichten wollten, haben wir die Möglichkeiten von Java zur Lösung dieser Probleme untersucht [2, 5]. Wir haben dazu mit Java eine Client/Server-Anwendung auf der Basis von RMI (Remote Method Invocation) und JDBC (Java Database Connectivity) entwickelt, welche zunächst die umfangreichste Benutzeroberfläche, nämlich die des Administrators zur Pflege des Klassifikationsschemas, enorm vereinfacht.

Abbildung 4 zeigt auf der linken Seite die Umgebung des Administrators. Im WWW-Browser wird über HTTP vom Web-Server eine HTML-Seite mit eingebettetem Java-Applet geladen. Das Applet besitzt neben der grafischen Benutzeroberfläche einen Teil, welcher einen RMI-Client realisiert. Dieser RMI-Client kommuniziert über den entfernten Methodenaufruf mit einem Java RMI-Serverprozeß, welcher über JDBC auf die in der Oracle-Datenbank gespeicherten Daten und PL/SQL-Routinen zugreifen kann. In diesem Szenario besteht keine Verbindung zwischen dem Web-Server und der Datenbank. Der Web-Server liefert lediglich die HTML-Seite und das Applet zusammen mit allen benötigten weiteren Klassen.

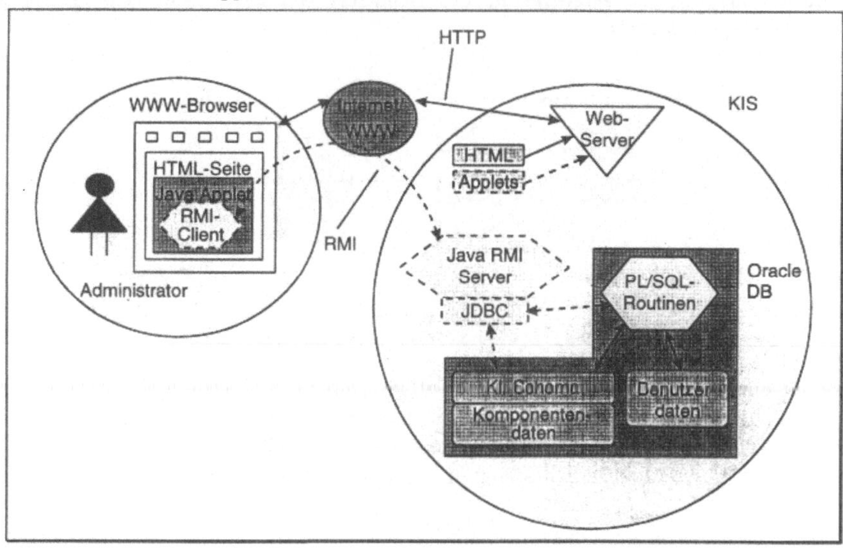

Abbildung 4: Java Client/Server mit RMI und JDBC

Um einen Vergleich des Aufwands (Implementierung, Performance) einer verteilten Java-Anwendung mit einer Einzelplatzversion durchführen zu können, haben wir zunächst eine Einzelplatzversion realisiert, welche anschließend auf der Basis von RMI zu einer Client/Server-Anwendung überarbeitet wurde. Auch mit der nicht-verteilten Version kann jedoch mit geeigneten JDBC-Treibern auf eine entfernte Datenbank zugegriffen werden.

Das Applet mit entferntem Zugriff über JDBC ist in Netscape 4.03 und HotJava lauffähig. Das Java-Applet im Client/Server-Szenario mit RMI konnte bisher lediglich im Appletviewer (JDK) evaluiert werden. Dies liegt an der zur Zeit noch unzureichenden Unterstützung der Browser hinsichtlich RMI und JDK1.1 allgemein.

Uns hat der Einfluß sowohl der Verteilung als auch der verschiedenen JDBC-Teiber auf die Performanz interessiert. Herausgestellt hat sich, daß im Client/Server-System durch RMI erwartungsgemäß eine erhöhte Bearbeitungszeit entsteht. Diese wächst linear mit der Menge der übertragenen Daten (bei unseren Messungen von 2 bis 150 KBytes im Mittel mit einer Rate von 16 KBytes/sec), läßt aber immer noch eine akzeptable Interaktivität zu. SUN hat die Menge der JDBC-Treiber in vier Kategorien eingeteilt, welche sich sowohl durch den Ort der Installation und die Kommunikation zur Datenbank als auch durch die zusätzlich zu Java eingesetzten Programmiersprachen unterscheiden. Es hat sich gezeigt, daß die beiden getesteten Treiber (Typ-2 bzw. Typ-4) für die Oracle-Datenbank unterschiedliche Stärken besitzen. Bei größeren Datenmengen zeigte der ("nur Java") Typ 4-Treiber schnellere Übertragungsraten, während bei geringeren Datenmengen der Typ 2-Treiber, schneller war. Neben den Installationsmöglichkeiten —Typ-2-Treiber können beispielsweise nicht in Applets verwendet werden— ist es also sinnvoll, die Performanz der vorhandenen Treiber miteinander zu vergleichen und bei der Wahl des Treibers die voraussichtlich zu übertragenden Datenmengen zu berücksichtigen. Mit den Typ-4 (und auch Typ-3) JDBC-Treibern ist es jederzeit möglich, von einem entfernten Rechner auf eine Datenbank zuzugreifen. An dieser Stelle muß Aufwand und Nutzen der Erstellung einer 3-Schichten-Architektur auf der Basis von RMI auch im Hinblick auf die Lastenverteilung und die Ladezeit beim Client abgewogen werden.

3.3 Nutzen von Internet-Technologien

Ein weiterer wichtiger Aspekt ist die Integration in das Internet. Das Informationssystem soll nicht nur im Internet verfügbar sein, es kann und muß auch im Bezug zu anderen Informationsquellen im Internet stehen. So können zu einer Software-Komponente Verweise (Links) auf beliebige zulässige Quellen im Internet, repräsentiert durch URLs (Uniform Resource Locators), angegeben werden. Bisher sind dies beispielsweise Verweise auf Web-Server von Herstellern und Anbietern, Verweise auf Produktbeschreibungsseiten, Verweise auf weitere Informationsquellen oder Verweise auf Seiten zum kostenlosen Download von Produkten, Trial- oder Demo-Versionen oder sogar zum Online-Kauf. Natürlich können auch die E-Mail-Adressen der Bezugsquellen sowie der Ansprechpartner zum Erfahrungsaustausch innerhalb REGINA angegeben werden. Erfahrungsaustausch wird aber nicht nur innerhalb der Gruppe unterstützt. Durch Verweise auf verschiedene für eine Komponente oder ein Anwendungsgebiet relevante *Newsgroups* kann der Entwickler direkt auf Erfahrungen von anderen Entwicklern weltweit zurückgreifen, indem er die Diskussionen verfolgt oder sich selbst beteiligt und Kontakte knüpft.

Durch diese Referenzen können sämtliche Informationen zentral gepflegt und bereitgestellt werden, im Gegensatz zu den Komponenten selbst (Quelltexte, Bibliotheken, Dokumentation, etc.), welche physikalisch verteilt jeweils bei den Herstellern oder Anbietern gepflegt werden. Wichtig ist dabei die Nutzung der bei den Anbietern bereits vorhandenen Infrastrukturen (Web-, FTP-Server, E-Mail), so daß dort keine neuen Installationen benötigt werden.

Auch die verschiedenen bei den Entwicklern vorhandenen WWW-Browser werden berücksichtigt. Neben der Java-Benutzerschnittstelle wird auch weiterhin für Browser, die Java nicht unterstützen, der Zugang über HTML-Seiten ermöglicht.

Das Hypertext Transfer Protocol kann außerdem genutzt werden, um den Zugang zu HTML-Seiten auf registrierte Benutzer zu beschränken. Spezielle Protokollelemente erlauben den Zugang zu speziellen HTML-Seiten nur, wenn ein gültiger Benutzername mit Paßwort vorhanden ist. Der WWW-Browser fragt dazu Benutzername/Paßwort ab und der Web-Server überprüft die Zulässigkeit.

4 Verwandte Arbeiten

Die Mehrzahl der bestehenden Ansätze beschäftigt sich mit Software-Wiederverwendungs-bibliotheken innerhalb eines Unternehmens. Eine detaillierte Übersicht zu prinzipiellen Möglichkeiten zur Repräsentation von wiederverwendbaren Software-Bausteinen liefert [9].

Ähnliche Ansätze zur Software-Klassifikation sind die facettierte Klassifikation [12] und die merkmalsorientierte Klassifikation [7]. Jedoch unterstützt die facettierte Klassifikation, welche auch in dem umfangreichen Wiederverwendungsprojekt REBOOT [10] eingesetzt wurde, keine Beschreibungstexte. Weiterhin können Komponenten nicht auf verschiedenen Granularitätsstufen beschrieben werden, bei REBOOT werden Komponenten bespielsweise lediglich auf Klassenebene betrachtet. Bei der merkmalsorientierten Klassifikation wird die Klassifikationshierarchie auf eine andere Art aufgebaut und ist insbesondere nicht auf verschiedene Zugriffsmechanismen (Suche/Navigation) ausgerichtet. Beide Ansätze unterstützen jedoch -im Gegensatz zu unserer derzeitigen Implementierung- Ähnlichkeiten zwischen Termen, so daß wenn kein Ergebnis zu einer Anfrage gefunden wurde, ein ähnliches Ergebnis geliefert wird.

Andere Systeme, die sich mit der Integration von Software-Wiederverwendung und Internet-Technologien beschäftigen sind ASSET [1] und MOREplus [14]. ASSET beschäftigt sich mit wiederverwendbaren Dokumenten ganz allgemein (Standards, Werkzeuge, Dokumentationen, aber auch Klassenbibliotheken) und ist weiterhin nicht auf registrierte Benutzer ausgerichtet. Bei dem kommerziellen Produkt MOREplus handelt es sich um den insgesamt ähnlichsten Ansatz im Vergleich zu dem vorgestellten Komponenten-Informationssystem.

Das System JAVA REPOSITORY [8] verwendet die gleiche Technologie für die Datenbank und den Web-Server wie der hier vorgestellte Ansatz. Im Gegensatz zu der Idee des KIS handelt es sich dabei jedoch nicht um einen kooperativen Ansatz. Über das Java Repository werden Java-Komponenten direkt auf der Basis von Electronic Cash verkauft. Die Beschreibung der Komponenten ist relativ kurz und basiert nicht auf einem hierarchischen Klassifikations-schema; es gibt acht Kategorien und verschiedene Schlüsselworte, denen eine Komponente zugeordnet werden kann. Technische Informationen (z.B. unterstütztes JDK, getestete Platt-formen) werden nicht angeboten.

5 Zusammenfassung und Ausblick

Vorgestellt wurden die Konzepte und die Realisierung des Komponenten-Informationssystems, das im Rahmen des RSB-Projektes entwickelt wird. Dieses Projekt findet in Kooperation mit neun Mitgliedern des Regionalen Industrie-Clubs Informatik Aachen statt. Ziel ist die Förderung der Kooperation bei der Wiederverwendung von Software zur Entwicklung neuer Software-Systeme.

Wir haben gezeigt, wie man durch Einsatz von Internet-Technologien eine Kooperation zwischen verschiedenen Software-Herstellern unterstützen kann. Wichtige Aspekte dabei waren neben der Berücksichtigung der bei den einzelnen Partner vorhandenen Infrastrukturen auch die Möglichkeiten zur Kommunikation und zum Erfahrungsaustausch zwischen den Projektpartnern, aber auch mit anderen Entwicklern weltweit.

Wir haben gezeigt, wie ein solches System durch die Kombination verschiedener Technologien realisiert wurde. Der Datenbestand, welcher in einer relationalen (Oracle) Datenbank auf der Basis eines neu entwickelten Software-Klassifikationsschemas abgespeichert wird, wird auf zwei verschiedene, jedoch kombinierbare Arten im Internet verfügbar gemacht. Eine Möglichkeit ist die dynamische Generierung der HTML-Benutzeroberfläche durch den Anschluß eines Web-Servers über einen "Web Agent" an die Datenbank. Dieser Ansatz ist jedoch für komplexe Oberflächen wenig geeignet. Als Alternative wurde ein Konzept für eine benutzerfreundlichere Client/Server-Anwendung auf der Basis von Java RMI und JDBC erläutert. Einflußfaktoren auf die Performanz dieses Client/Server-Systems wurden aufgezeigt. Zur Zeit arbeiten wir an weiteren Java-Applikationen und Applets, welche nach dem oben beschriebenen Konzept realisiert werden, um die Suche und Navigation benutzerfreundlicher zu gestalten.

Es hat sich herausgestellt, daß das System von den Benutzern zwar prinzipiell für nützlich befunden und akzeptiert wird, jedoch der Umfang der enthaltenen Informationen nicht ausreichend ist. Daher arbeiten wir an einem Werkzeug, welches unter Nutzung der vorhandenen Internet-Suchmaschinen weitere Komponenten findet, deren Beschreibungen im Internet verfügbar sind. Für den aufwendigen Prozeß der Eingabe von Komponentendaten wird ferner ein interaktives Werkzeug entwickelt, welches Text- oder HTML-Dateien auswertet, die beschreibende Texte (Produktbeschreibung, Dokumentation, ...) enthalten.

Literaturverzeichnis

[1] Science Applications International Corporation (SAIC): ASSET - Asset Source for Software Engineering Technology. Juni 1997, http://www.asset.com/WSRD/background.html

[2] M.A. Al-Sayed Ali: Einsatz von Java zur Entwicklung von Client/Server-Systemen am Beispiel einer Datenbankanwendung im World Wide Web. Diplomarbeit des Lehrstuhls für Informatik III an der RWTH Aachen, Februar 1998.

[3] A. Behle, A. Deparade, P. Klein, M. Nagl: Specific Application Environments. In M. Nagl (Hrsg.): Building Tightly Integrated Software Development Environments: The IPSEN Approach. Lecture Notes in Computer Science, Vol. 1170, Berlin, Germany, 1996: Springer-Verlag. S. 592-605.

[4] A. Behle: An Internet-based Information System for Cooperative Software Reuse. erscheint in Proc. of 5th International Conference on Software Reuse, Victoria, BC, Canada, Juni 1998.

[5] A. Behle, K. Cremer: Einsatz und Lehre von Java zur Entwicklung verteilter Applikationen. In: Tagungsband Smalltalk und Java in Industrie und Ausbildung, Erfurt, September 1997. Ilmenau, 1997: TU Ilmenau. S. 108-113.

[6] A. Behle: Ein dynamisches, benutzerabhängiges Informationssystem im World Wide Web. EMISA-Fachgruppentreffen Okt. 1996 „Informationsserver für das Internet - Anforderungen, Konzepte, Methoden", Aachen, EMISA Forum 1/1997, S. 75-79.

[7] J. Börstler: Feature-Oriented Classification for Software Reuse. In: Proceedings of the 7th International Conference on Software Engineering and Knowledge (SEKE '95), Rockville, ML, USA, Juni 1995, S. 204-211.

[8] P. Buxmann, W. König, F. Rose: The Java Repository - An Electronic Intermediary for Java Resources. 7th Annual Conference of the International Information Management Association (IIMA), Estes Park, Colorado, Dezember 1996, http://java.wiwi.uni-frankfurt.de/papers/JavaRepository_eng.html.

[9] W.B. Frakes, P.B. Gandel: Representing Reusable Software. Information and Software Technology 32(10), 1990, S. 641-664.

[10] E.-A. Karlsson: Software Reuse - A Holistic Approach. Chichester, England, 1995: John Wiley & Sons.

[11] M. Matsumoto: SEA/I: Systems Engineer's Arms for Industrialized Production and Support of Application Programs. In: Proceedings of 6th International Conference on Software Engineering, Tokio, September 1982.

[12] R. Prieto-Díaz: Implementing Faceted Classification for Software Reuse. In: Communications of the ACM 34(5), Mai 1991, S. 88-97.

[13] W. Tracz: Confessions of a Used Program Salesman - Institutionalizing Software Reuse. Reading, Massachusetts, 1995: Addison Wesley.

[14] D. Trump: Using the WWW and the Internet to Support Corporate Reuse. In: Proceedings 8th Annual Workshop on Institutionalizing Software Reuse (WISR8). Ohio State University, Columbus, Ohio, USA, März 23-26, 1997.

[15] A. Zendler: Internationale Erfahrungsberichte zur Software-Wiederverwendung. Forschungsinstitut für Angewandte Software-Technologie, Bericht Nr. 95-03, München, Februar 1995, http://www.fast.de/Publikationen/Forschungsberichte/1995/.

Autorin

Dipl.-Inform. Anita Behle
Lehrstuhl für Informatik III
RWTH Aachen
52056 Aachen
E-Mail: behle@informatik.rwth-aachen.de

Konzeption und Entwicklung einer Internet-basierten Zeitung am Beispiel von EIWIZ, der *El*ektronischen *W*irtschafts*I*nformatik-Zeitung

Klement J. Fellner, Susanne Patig, Claus Rautenstrauch
Institut für Technische und Betriebliche Informationssysteme
Otto-von-Guericke-Universität Magdeburg

Zusammenfassung

In diesem Beitrag wird die Implementierung einer elektronischen Zeitung beschrieben, die zusätzlich zu vorhandenen statischen Web-Seiten tagesaktuelle Informationen bereitstellt.

Der potentielle Nutzen, den das WWW, als Informationsmedium mit stetig wachsender Bedeutung, erbringt, wird oftmals durch mangelnde Aktualität der angebotenen Informationen beeinträchtigt. Verursacht durch den hohen Administrationsaufwand, den z. B. statische HTML-Seiten mit sich bringen, werden Informationen im Web nur in unregelmäßigen Abständen aktualisiert.

Bei der Konzeption der in Java implementierten elektronischen Zeitung wurde daher besonderer Wert auf eine effiziente und einfache Administration der Inhalte gelegt. Die Publikation aktueller Information erfolgt dabei online über spezielle Module der elektronischen Zeitung, welche auch die präsentationsgerechte Formatierung übernimmt.

1 Elektronische Zeitungen und Zeitschriften

Elektronische Zeitungen (Online-Zeitungen) und elektronische Zeitschriften (E-Journals) gewinnen zunehmend Bedeutung im Informationsangebot des Internet. Anbieter elektronischer Zeitungen und Zeitschriften sind zum größten Teil diejenigen, die bereits konventionelle Nachrichtenmagazine, Tageszeitungen oder auch wissenschaftliche Journale auf dem Markt anbieten. Dabei übertragen Verlage/Herausgeber teilweise die Inhalte konventioneller Zeitungen und Zeitschriften mehr oder weniger unverändert auf Web-Seiten (ein Beispiel hierfür ist die Zeitschrift „Wirtschaftsinformatik" [14]), andere schöpfen informationelle Mehrwerte aus, indem sie Ihren Internet-Auftritt mit Links, Interaktionsmöglichkeiten u.ä. ergänzen (z.B. die Nachrichtenmagazine „Focus" [6] und „Die Zeit" [5]). Zeitungsverlage sehen das Internet-Angebot eher als Ergänzung zu ihren Druckerzeugnissen an, mit denen die Möglichkeiten elektronischen Publizierens eruiert werden und der Verkauf der Druckerzeugnisse gefördert werden soll.

Eine Ausnahme sind bisher noch Zeitungen und Zeitschriften, die ausschließlich im Internet verfügbar sind. Ziele derartiger Zeitungen und Zeitschriften sind vor allem, dem Veröffentlichungswildwuchs im Internet in der Form zu begegnen, daß auch hier wissenschaftliche Qualitätssicherungsmechanismen Anwendung finden, sowie die Gewährleistung der persistenten Verfügbarkeit von Veröffentlichungen, was Grundlage für die Zitierfähigkeit von ausschließlich im Internet veröffentlichten Artikeln ist. Beispiel für eine solche Zeitung ist der „Review on Information Science" [12], ein internationales referiertes E-Journal, das ausschließlich in elektronischer Form verfügbar ist [1].

Auch die in diesem Artikel behandelte Online-Zeitung ElWIZ, die **Elektronische WirtschaftsInformatik-Zeitung**, ist ausschließlich in elektronischer Form verfügbar. Das strategische Ziel von ElWIZ ist, Studierenden und anderen an der Arbeitsgruppe[1] Wirtschaftsinformatik der Otto-von-Guericke-Universität Magdeburg interessierte Personen tagesaktuell Informationen zu liefern. Gleichzeitig wird ein informelles Forum für den Informationsaustausch zwischen Studierenden und Mitarbeitern der Arbeitsgruppe geschaffen.

2 Informationelle Mehrwerte von Online-Zeitungen und E-Journals

Die Vorteile von Online-Zeitungen und E-Journals liegen in der Möglichkeit, mit der Präsentation im Internet informationelle Mehrwerte auszuschöpfen, die mit konventionellen Druckerzeugnissen nicht möglich sind. Die Theorie der informationellen Mehrwerte [9, S. 80ff] wird hier als Bezugsrahmen für die qualitative Bewertung elektronischer Dienstleistungen herangezogen. Die für die Bewertung von Online-Zeitungen und E-Journals relevanten Mehrwerte werden im folgenden dargestellt:

- *Komparative Mehrwerte*: In Online-Zeitungen und E-Journals lassen sich Informationen im Vergleich zu konventionellen Druckerzeugnissen adäquat darstellen. Beispielsweise können hier Tondokumente, Animationen und Videosequenzen abgespielt werden.

- *Strategische Mehrwerte*: Informationen stehen prinzipiell weltweit zur Verfügung. Anders ausgedrückt: Die „Auflage" einer Online-Zeitung oder eines E-Journals ist aus Anbietersicht unendlich groß.

- *Innovative Mehrwerte*: Leser können Artikel, sofern dies von den Autoren zugelassen ist, zur Weiterverarbeitung herunterladen. Weiterhin ist die Möglichkeit zur direkten Interaktion mit der Redaktion via E-Mail möglich, und Leserbriefe können ebenfalls per E-Mail versendet werden. In Online-Zeitungen oder E-Journals können dann Artikel, Leserbriefe hierzu, Leserbriefe mit Kommentaren zu Leserbriefen usw. ähnlich wie in Newsgroups strukturiert angeboten werden. Außerdem können über Links Informationsangebote (wie z.B. im Internet veröffentlichte Stellenangebote) leicht integriert werden.

---

[1] Der Begriff „Arbeitsgruppe" wird in Magdeburg als Synonym für Lehrstuhl verwendet.

- *Mehrwerte mit Effizienzwirkung*[2]: Online-Zeitungen und E-Journals lassen sich verhältnismäßig kostengünstig produzieren, da die gesamten Druck- und Distributionskosten entfallen. Weiterhin lassen sich mit Online-Zeitungen und E-Journals Zeitvorteile erreichen, da zwischen der Verfügbarkeit von Beiträgen und deren Veröffentlichung nur noch Zeitaufwand für redaktionelle Bearbeitung und Satz, nicht jedoch für Druck und Distribution, anfällt. Bei Online-Zeitungen und E-Journals entfällt damit auch die prinzipielle Terminbindung für Ausgaben (permanente Aktualisierung).

- *Mehrwerte mit Effektivitätswirkung*[3]: Unter der Voraussetzung, daß die Zielgruppe vornehmlich Internet-Benutzer sind, ist bei Online-Zeitungen und E-Journals eine wesentlich höhere Effektivität bei der Erreichung der potentiellen Leser als bei konventionellen Zeitungen und Zeitschriften gegeben. An die Stelle konventioneller Distributionskanäle treten Einträge in Internet-Suchdiensten, anhand derer Internet-Benutzer gezielt auf das Informationsangebot von Online-Zeitungen und E-Journals geleitet werden können. Auch bei der Gestaltung des Informationsangebotes lassen sich Effektivitätsvorteile erzielen. So können durch eine flexible Informationsaufbereitung, die nicht an Ausgabetermine gebunden ist, Informationen genau solange im Angebot verbleiben, wie sie aus Sicht des Publizierenden relevant sind. Jedem Artikel wird hier ein Gültigkeitszeitraum zugeordnet, der verkürzt oder verlängert werden kann. Nach Ablauf des Gültigkeitszeitraums wird ein Artikel archiviert und gehört nicht mehr zum originären Angebot. Eine retrospektiv ausgerichtete Wartung des Informationsangebots, wie sie bei konventionellen Internet-Seiten üblich ist, wird dadurch überflüssig.

3 ElWIZ im Überblick

Betrachtet man vorhandene Online-Zeitungen und E-Journals, so heben sich diese von Internet-Präsentationen mit statischen HTML-Seiten durch hohe Interaktivität und einen redaktionellen Rahmen ab, der sowohl das Design der Seiten, die Strukturierung der Informationen (z.B. in Form von Rubriken) und die inhaltliche Qualitätssicherung umfaßt [1, S. 146f]. Vor diesem Hintergrund fiel die Entscheidung, Informationen der Arbeitsgruppe Wirtschaftsinformatik der Otto-von-Guericke-Universität Magdeburg nicht nur statisch, in Form periodisch gewarteter HTML-Seiten, sondern zusätzlich durch die Online-Zeitung ElWIZ zu präsentieren.

---

[2] Unter Effizienzwirkung verstehen wir in diesem Zusammenhang den wirtschaftlichen Einsatz von Ressourcen [9], S. 91; [13], S. 287).

[3] Unter Effektivitätswirkung wird hier eine verbesserte Zielerreichung durch den Einsatz von Ressourcen verstanden [9], S. 91; [13], S. 287).

Mit ElWIZ soll das vorhandene statische Informationsangebot, das Informationen zu Mitarbeitern, dem Lehrangebot, den laufenden Forschungsvorhaben und Publikationen beinhaltet, um aktuelle Informationen zu Aktivitäten der Arbeitsgruppe Wirtschaftsinformatik, der Fakultäten für Informatik und Wirtschaftswissenschaft sowie sonstige interessante Entwicklungen in der Wirtschaftsinformatik-Community erweitert werden. Die Kunden dieses Informationssystems sind primär die Studenten der Wirtschaftsinformatik, die durch Lehrveranstaltungen, Praktika und Diplomarbeiten Kontakt zur Arbeitsgruppe haben. Aber auch Fakultäten, Institute und Lehrstühle anderer Universitäten sowie Unternehmen gehören zur Zielgruppe der Online-Zeitung der Arbeitsgruppe Wirtschaftsinformatik. Die einzelnen Kunden der Online-Zeitung sind zwar in ihrem Informationsbedarf sehr unterschiedlich, benötigen jedoch jederzeit aktuelle Informationen in ansprechender Form.

Wichtig ist auch, daß trotz eines einheitlichen Erscheinungsbildes von ElWIZ sichergestellt wird, daß aktuelle Informationen auf den ersten Blick erkennbar sind. Hierfür wird auf der Homepage nicht nur das letzte Änderungsdatum, sondern auch der aktuellste Artikel - unabhängig von der Zugehörigkeit zu einem Inhalt (s. Abbildung 1) - direkt angezeigt.

4 Realisierung von ElWIZ

Ein gravierender Nachteil der Informationsaufbereitung mit statischen HTML-Dokumenten ist der hohe Administrationsaufwand bei Inhalts- und Layoutänderungen, da jedes Dokument einzeln bearbeitet und angepaßt werden muß dazu [8, S. 75]. Der Administrationsaufwand bei inhaltlichen Änderungen kann durch eine dynamische Generierung der Dokumente mit Hilfe von CGI (Common Gateway Interface) und der Speicherung der Information in einer Datenbank reduziert werden [2, S. 4ff]. Aber auch hier sind die Meta-Informationen der übermittelten Dokumente, wie Layout, Inhaltsstruktur, etc. statisch in Form von HTML-Dateien im Filesystem des Web-Servers abgelegt. Die Interaktion mit dem Benutzer ist zudem auf die Möglichkeiten, die HTML-Formulare bieten, beschränkt. CGI-Programme können mit jeder auf dem Entwicklungssystem verfügbaren Programmiersprache, z. B. C, Perl unter Unix oder Windows, Visual Basic unter Windows oder Applescript unter MacIntosh, erstellt werden. Jedoch sind CGI-Programme durch den Zugriff auf Umgebungsvariablen des Betriebssystems generell plattformabhängig sowie durch das Fehlen von Schnittstellenstandards für den Datenbankzugriff auch datenbankabhängig [4, S. 27]; [10, S. 151ff].

Die genannten Nachteile von CGI-Programmen führten in Verbindung mit der angestrebten Interaktivität und Flexibilität der Online-Zeitung zur Wahl von Java als Programmiersprache sowohl für den Client als auch für die Realisierung der Datenbankzugriffe. Zudem spricht für Java die hohe Akzeptanz bei Anwendern und die Unterstützung durch alle am Markt verfügbaren Web-Browser, sowie die Plattformunabhängigkeit. Der Zugriff auf die im RDBMS (Relational Database Management System) gespeicherten Daten wurde über die in Java integrierte Programmierschnittstelle JDBC (Java Database Connectivity) realisiert. Das

Datenmanagement ist damit generell mit jedem Datenbankmanagementsystem (DBMS) möglich, sofern dafür JDBC-Treiber angeboten werden und die verwendeten DML (Data Manipulation Language) und DDL (Data Definition Language) Befehle dem ANSI SQL92-Standard entsprechen [4, S. 27ff].

5 Informationsstruktur

Als durch ElWIZ tagesaktuell darzustellende Informationen wurden in der Analysephase folgenden Bereiche ermittelt (s. Abbildung 1):

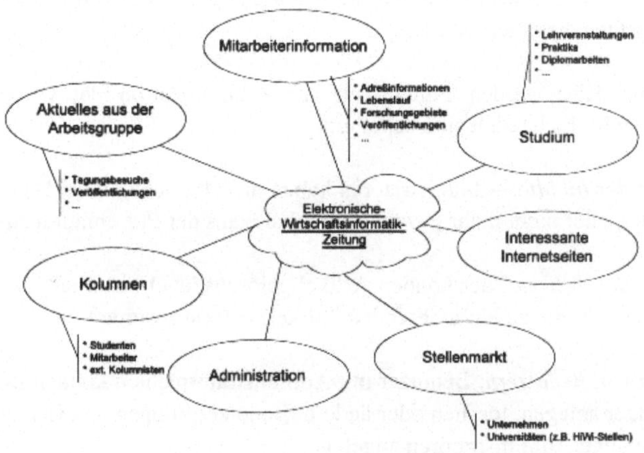

Abbildung 1: Inhalte der Online-Zeitung

- *Aktuelles aus der Arbeitsgruppe*: Hier werden aktuelle Termine und Tätigkeiten der Mitglieder der Arbeitsgruppe publiziert. Dazu gehören z.B. Informationen der Art „Mitarbeiter X trägt auf Tagung in Y vor" oder „Mitarbeiter Z verteidigt am dd.mm.yyyy seine Dissertation".

- *Mitarbeiterinformation*: Die Mitarbeiter der Arbeitsgruppe werden mit Daten zu ihrer Person präsentiert. Zusätzlich werden Forschungsschwerpunkte und Veröffentlichungslisten personenbezogen angegeben.

- *Studium (Wirtschaftsinformatik)*: Die Studenten bekommen aktuelle Informationen zu angebotenen Lehrveranstaltungen, Praktikums- und Diplomarbeitsthemen sowie zur Diplomprüfungs- und Studienordnung. Zu den tagesaktuellen Informationen gehören insbesondere Raum- und Terminänderungen sowie neue oder geänderte Begleitmaterialien zu Lehrveranstaltungen.

- *Interessante Internetseiten*: Verweise zu für die Wirtschaftsinformatik interessanten Web-Seiten können in der Online-Zeitung zentral verwaltet werden (z. B. Gesellschaft für Informatik (GI), Wirtschaftsinformatiklehrstühle im deutschsprachigen Raum).

- *Stellenmarkt*: Ein Dienst, der für Studenten und Absolventen interessante Stellenangebote aus Forschung und Wirtschaft enthält.

- *Kolumnen*: Bei bekundetem Interesse können Online-Kolumnen eingerichtet werden, in denen ausgewählte Personen ihre Meinung zu einem Thema veröffentlichen können. Hier könnten z.B. Artikel zu neuen Rahmenstudienordnungen oder Kommentare zu aktuellen Hochschulgesetzen diskutiert werden.

- *Administration*: Hier werden Funktionen zur Administration der Online-Zeitung zur Verfügung gestellt. Es handelt sich dabei um:

  - *Verwalten der Inhalte*: Administratoren haben die Möglichkeit, bei Bedarf neue Inhalte einzurichten oder nicht mehr gewünschte Inhalte aus der Präsentation zu entfernen.

  - *Verwalten der Artikel*: Hier können Artikel gelöscht (auch innerhalb des Archivs) oder als verfallen (für die zu archivierenden Artikel) markiert werden.

  - *Verwalten von Benutzern*: Benutzer mit Administratorrechten können Benutzergruppen und Benutzer anlegen, löschen oder ändern. Benutzergruppen und Benutzer werden nur für Autoren und Administratoren angelegt.

Zu den Bereichen „Aktuelles aus der Arbeitsgruppe", „Interessante Internetseiten", „Stellenmarkt" und den Kolumnen können online von jedem Leser der Online-Zeitung Stellungnahmen, sog. Leserbriefe, geschrieben werden.

5.1 Benutzermodell

Die Benutzer der Online-Zeitung (s. Abbildung 2) können wie folgt eingeteilt werden (zur Benutzerdefinition [11, S. 63ff]):

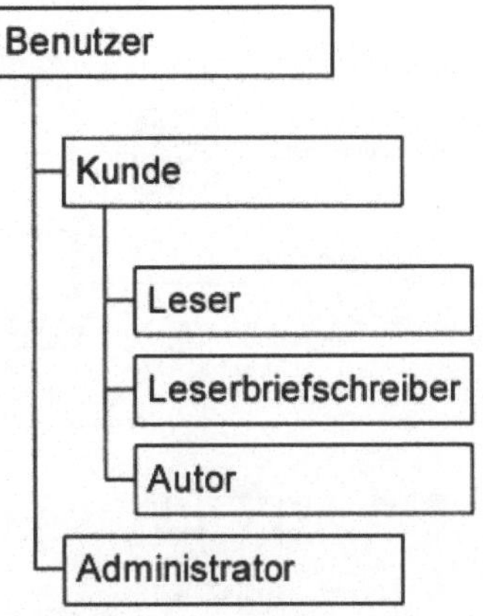

Abbildung 2: Benutzerklassifizierung

- *Leser:* Als Leser werden diejenigen Kunden eingestuft, die keinerlei Artikel für die Online-Zeitung schreiben.

- *Leserbriefschreiber:* Kunden haben grundsätzlich die Möglichkeit, durch das Schreiben von Leserbriefen auf in der Online-Zeitung publizierte Artikel zu reagieren. Diese Meinungsäußerung wird online durchgeführt und steht weiteren Kunden sofort nach dem Absenden durch den Leserbriefschreiber in der Online-Zeitung zur Verfügung. Leserbriefe werden dabei zusätzlich dem jeweiligen Redakteur (Inhaltsverantwortlichen) per E-Mail zugestellt (s. Abbildung 3). Der Redakteur greift in den Publikationsprozeß eines Leserbriefes nur ein, wenn der Inhalt aus gesetzlichen oder sittlichen Gründen abzulehnen ist (Darstellung in Abbildung 3 als Kontrollfluß).

- *Autoren:* Die Gruppe der Autoren umfaßt diejenigen Personen, die Artikel zu einzelnen Inhalten der Online-Zeitung (s. Abbildung 1) liefern. Autoren besitzen im Gegensatz zu Leserbriefschreibern das Recht in ihrem Zuständigkeitsbereich Artikel, sowie Leserbriefe zu löschen oder durch ändern der Verfallsdaten für die Archivierung einzustellen. Außerdem treten die Autoren der Kolumnen gegenüber Leserbriefschreibern als Redakteure auf, erhalten somit alle Leserbriefe per E-Mail zugestellt.

- *Administratoren:* Für jeden publizierten Inhalt (s. Abbildung 6) in der Online-Zeitung ist eine Person als Administrator zuständig. Diese Zuständigkeit erstreckt sich bei mindestens einem Benutzer auf die gesamte Online-Zeitung (Chefredakteur). Administratoren können z. B. die Auswahl der zu archivierenden Artikel beeinflussen oder neue Autoren definieren.

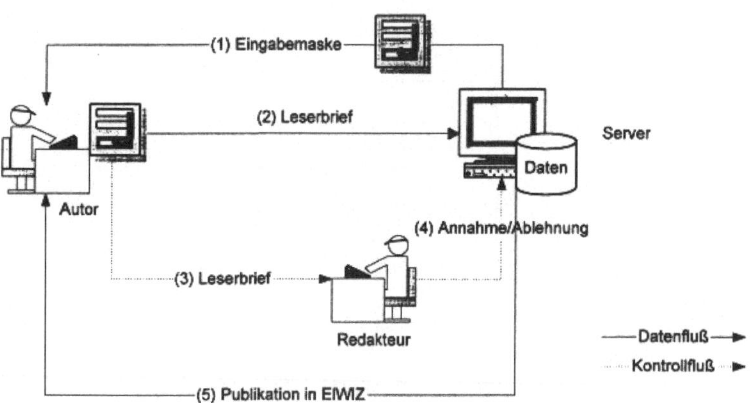

Abbildung 3: Schreiben und Beurteilen eines Leserbriefs

Benutzer können mehreren der nicht disjunkten Benutzerklassen zugeordnet werden. Ein Autor tritt z. B. auch als Leserbriefschreiber auf, wenn er einen Leserbrief zu einem Artikel eines anderen Autors veröffentlicht.

Für den Bereich der Lehrveranstaltungen ist i. d. R. der jeweilige Dozent für den Inhalt der Informationen auf den entsprechenden Seiten verantwortlich. Meinungen und Informationen der Hörer der Lehrveranstaltungen können hier nicht direkt mittels Leserbrief in der Online-Zeitung publiziert werden, sondern sind per E-Mail an den Dozenten zu senden. Hörer haben jedoch die Möglichkeit, über eigene Kolumnen zu einzelnen Lehrveranstaltungen Informationen auszutauschen. Der Bereich „Lehrveranstaltungen" ist gegenüber diesen Kolumnen als Dienstleistung der Dozenten an die Hörer anzusehen.

Mitarbeiter der Arbeitsgruppe sind prinzipiell nur noch für den Inhalt der Online-Zeitung zuständig, nicht wie bisher auch für den Aufbau der publizierten Webseiten. Sie übernehmen somit rein redaktionelle Tätigkeiten, bei denen sie vom Administrationsteil Online-Zeitung unterstützt werden. Dabei müssen Mitarbeiter nicht zwangsläufig Administratoren sein. Mitarbeiter, die z. B. nur für ihre Lehrveranstaltungen in der Online-Zeitung publizieren, benötigen keinerlei Administrationsrechte für andere Inhalte.

5.2 Benutzeroberflächengestaltung

Abbildung 4: Benutzeroberfläche der Online-Zeitung[4]

Mit dem Einsatz von Java stehen dem Entwickler umfangreichere Gestaltungsmöglichkeiten zur Verfügung als bei der Verwendung von HTML-Dokumenten. Java stellt für die Gestaltung der Benutzeroberfläche eine eigene Klassenbibliothek (Abstract Window Toolkit (AWT)) zur Verfügung [3, S. 322ff]. Die Präsentation der Inhalte kann damit über die übliche Form der Eingabemasken in HTML-Dokumenten hinausgehen. So ist die Möglichkeit der Verwendung von Bild und Ton bereits in der Standard-API (Application Programming Interface) vorgesehen. Des weiteren ist die Benutzerinteraktion durch ein direktes Feedback gekennzeichnet, womit z. B. eine sofortige Fehlerbehandlung bei Eingabe der Daten am Client ermöglicht wird und nicht wie bei Verwendung von HTML-Formularen eine Fehlermeldung bei Falscheingabe erst nach der Prüfung durch das CGI-Programm am Server erfolgt (zur Benutzerinteraktion [7, S. 101ff]). Außerdem können alle von der Programmierung grafischer Benutzeroberflächen bekannten Interaktionsobjekte, wie Fenster, Grafiken, Popup-Menüs, etc., zur Gestaltung der Benutzeroberfläche und somit zur Interaktion mit dem Benutzer verwendet werden.

Die Präsentationskomponente der Online-Zeitung wurde komplett in Java realisiert (s. Abbildung 4). HTML-Seiten dienen nur als Container, um das entsprechende Java-Applet am jeweiligen Client zu starten. Alle weiteren Navigationsschritte werden ausschließlich im Java-

---

[4] Die Abbildung zeigt die prototypische Benutzeroberfläche der Projektgruppe I (N. Appel, P. Krapp, A. Kreuzberg, B. Pieper, J. Reidemeister, A. Schmidt, T. Schmidt).

Applet ausgeführt. So werden die vorhandenen Inhalte aus der Datenbank ausgelesen und dem Benutzer in „Baumform" zur Auswahl angeboten. Bei Wahl eines Inhalts werden die darin enthaltenen Informationsobjekte (Artikel) in einer Schnellübersicht angezeigt, wobei die jeweils aktuellste Information komplett im Textfenster dargestellt wird.

Da jedoch nicht alle Kunden der Online-Zeitung Java-fähige Web-Browser einsetzen oder der Java-Interpreter im Browser unter Umständen aus Sicherheitsgründen o. ä. deaktiviert wurde, wird auch eine nicht auf Java basierende Version der Online-Zeitung angeboten. Diese HTML-basierte Version wird über einen Java-Server realisiert, der alle Informationen aus der Datenbank in HTML-konforme Informationen umwandelt und diese über den Web-Server dem Browser zur Verfügung stellt. Die HTML-basierte Implementation bietet gegenüber der Java-Version für Autoren eine eingeschränkte Funktionalität. Des weiteren sind im Vergleich zur Java-basierten Implementierung die Interaktionsmöglichkeiten auf Grund der Einschränkungen von HTML (zur Interaktion mit Online-Formularen [8, S. 228ff]) geringer.

6 Technische Realisierung

Neben der Benutzeroberfläche wurden auch sämtliche Datenbankzugriffe in Java realisiert. Verwendet wurde dazu ein Typ-4 JDBC-Treiber, der gemäß Spezifikation zu 100% in Java implementiert ist [4, S. 68]. Typ-4 JDBC-Treiber werden mit dem Applet vom entsprechenden Web-Server zum Client geladen und benötigen keine zusätzliche Datenbank-spezifische Schnittstellensoftware am Client. Durch die Verwendung der in Java implementierten JDBC-Treiber ist das Applet auf jedem Java-fähigen Client ausführbar. Um diese Prämisse der universellen Einsetzbarkeit aufrechterhalten zu können, wurde als Programmiersystem die Version 1.0.2 des JDK-API verwendet. Datenbanktreiber ist der zum verwendeten Oracle7 RDBMS verfügbare JDBC-Treiber.

6.1 Client/Server-Architektur

Abbildung 5 zeigt die verwendete Client/Server-Architektur. Die Online-Zeitung wurde, wie aus der Abbildung ersichtlich, in einer Zwei-Ebenen-Architektur implementiert. Auf dem Datenbankserver ist auch der Web-Server installiert, von dem die Applets auf den Client geladen werden. Dies ist aufgrund der Sicherheitseinschränkungen, die bei der Verwendung von Applets gelten, notwendig dazu [3, S. 173f]. Applets können demnach, sofern sie nicht signiert sind, nur mit dem Server eine Verbindung aufbauen, von dem sie auf den Client geladen wurden. Nach dem Laden des Applets wird eine SQL-Verbindung zum Datenbankserver aufgebaut, über die alle nachfolgenden SQL-Anweisungen sowie Ergebnismengen („ResultSets") übermittelt werden. Für die Validierung der Benutzerrechte der Autoren wird eine eigene Datenbankverbindung verwendet, die nur lesenden Zugriff auf die Datenbank erlaubt. Damit wird sichergestellt, daß nur Administratoren mit entsprechenden Zugriffsberechtigungen Autoren definieren können.

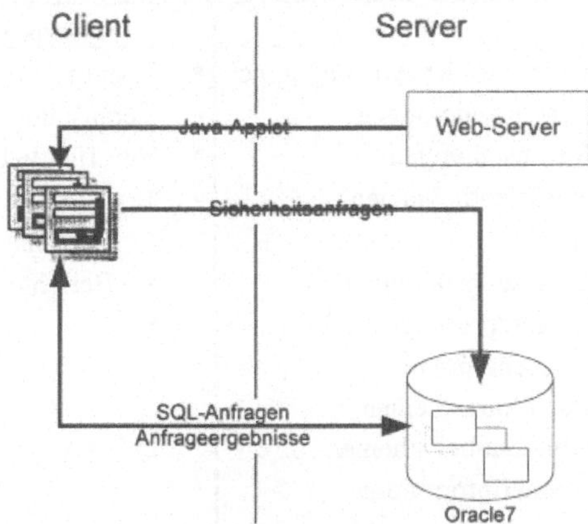

Abbildung 5: Zugriff auf den Datenserver

6.2 Implementierung der Benutzerrechte

Autoren der Online-Zeitung werden vom Administrator in einer eigenen Datenbanktabelle verwaltet. Die Datenbanktabelle ist für alle Kunden der Online-Zeitung gesperrt und wird nur beim Anmelden eines Autors mittels einer schreibgeschützten Verbindung zur Datenbank ausgelesen. Die Administration kann von Benutzern mit entsprechenden Administratorrechten über den Administrationsteil der Online-Zeitung oder aber direkt in der Datenbank erfolgen.

6.3 Artikel in der Online-Zeitung

Alle Veröffentlichungen in der Online-Zeitung werden als Artikel behandelt (s. Abbildung 7). Über die Zuordnung zu einem Inhalt und der Bestimmung des Artikeltyps werden die Artikel klassifiziert und für die Anzeige selektiert (s. Abbildung 6). Zusätzlich werden zu jedem Artikel das Erfassungsdatum sowie ein Verfallsdatum, das der Verfasser bestimmen kann, gespeichert. Bei der Auswahl der anzuzeigenden Artikel wird mit Hilfe des Standardverfallsdatums des Inhalts, der Erfassungs- und Verfallsdaten der einzelnen Artikel sowie des aktuellen Systemdatums ermittelt, welche Artikel in den jeweiligen Inhalten angezeigt und welche im Archiv abgelegt werden.

Inhalt	Artikeltypen

▪ Aktuelles aus der Arbeitsgruppe
  ▪ Terminänderungen
  ▪ Veranstaltungen
  ▪ Veröffentlichungen
▪ Studium
  ▪ Lehrveranstaltungen
  ▪ Diplomprüfungs- und
    Studienordnung
  ▪ Praktikumsthemen
  ▪ Diplomarbeitsthemen
▪ Mitarbeiterinformation
▪ Kolumnen (Beispiele)
  ▪ Mitarbeiter
  ▪ Studentenalltag
  ▪ Diplomarbeitsberichte
...
	▪ Termin
	▪ Veranstaltung
	▪ Veröffentlichung
	▪ Text
	  ▪ Kolumne
	  ▪ Berichte
	  ▪ ...

Abbildung 6: Inhalte und Artikeltypen in der Online-Zeitung

Damit wird sichergestellt, daß z. B. Termine, die bereits verfallen sind, nicht mehr angezeigt werden. So wird ein Artikel, der als Termin klassifiziert und für ein Ereignis am 15.10.1997 eingestellt wurde, am 16.10.1997 nicht mehr in der Online-Zeitung publiziert. Wird bei einem Artikel kein Verfallsdatum angegeben, greift das Standardverfallsdatum, das für jeden Inhalt definiert werden kann. Beispielsweise kann für die Inhalte, die eine Lehrveranstaltung betreffen, als generelles Verfallsdatum für alle Artikel das Semesterende eingegeben werden. Nach diesem zu spezifizierenden Datum wird folglich auch die Lehrveranstaltung aus der „aktuellen Ausgabe" der Online-Zeitung genommen, da keine zu publizierenden Artikel vorhanden sind. Alle für die Auswahl der zu publizierenden Artikel notwendigen Operationen (Datumsvergleiche) können direkt bei der Ausführung der Datenbankanfrage durchgeführt werden und erfordern somit keinerlei Berechnungen auf dem Client.

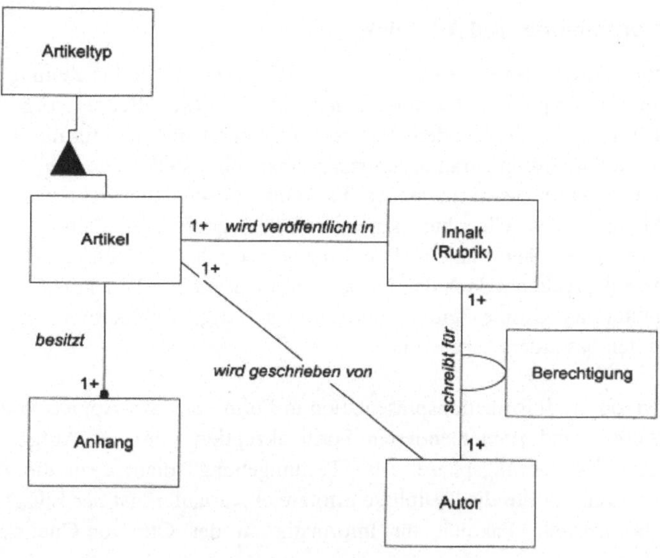

Abbildung 7: Objektschema von EIWIZ

6.4 Archiv

Das Archiv stellt die Sammlung aller Artikel dar, die nicht in der „aktuellen Ausgabe" der Online-Zeitung erscheinen. Archivierte Artikel können jedoch weiterhin abgerufen und gelesen werden. Die Artikel des Archivs werden in Form einer nach Datum geordneten Liste präsentiert. Die Auswahl und Anzeige der für einen Kunden interessanten Artikels erfolgt auf dieselbe Art und Weise, wie dies für „aktuelle" Artikel geschieht.

Zudem können jederzeit Artikel über den Administrationsteil (bis zu einem bestimmten Datum) aus dem Archiv gelöscht werden. Um eine lückenlose Datenhaltung zu gewährleisten, ist vor der Durchführung dieser Aktivität ein Backup der Daten durch den Datenbankadministrator durchzuführen.

6.5 Anhänge

Zu jedem Artikel können optional Anhänge in Form von Dateien oder Hyperlinks gespeichert werden, die dem Leser der Online-Zeitung zum Download oder zur Navigation zur Verfügung stehen. Bei Dateien enthält die Datenbank einen Verweis auf den Standort der Dateien am Server, der benutzt wird, um die gewünschten Dateien auf den Client zu laden.

6.6 Zusammenfassung und Ausblick

Wir haben eine Möglichkeit gezeigt, wie mit Hilfe einer Online-Zeitung aktuelle Informationen im Internet publiziert werden können. Die in diesem Beitrag vorgestellte Online-Zeitung bietet Vorteile für die Kunden, die besser informiert sind, und für die Bereitsteller der Informationen, deren Administrationsaufwand sinkt. Sie stellt eine eigenständige, neue Methode zur Informationsvermittlung dar. Es kann kritisch hinterfragt werden, inwieweit diese neue Methode den Charakter konventioneller, gedruckter Zeitungen bewahrt, da Ausgaben im herkömmlichen Sinne nicht existieren. Aktualität, Struktur und Zielstellung der Online-Zeitung entsprechen jedoch den Merkmalen traditioneller Druckmedien. Dabei ist die größere Aktualität der Online-Zeitung durch die sofortige Publikation ohne Druck- und Distributionszeiten besonders hervorzuheben.

Herauszufinden, ob die Informationspräsentation in Form von Java-Applets von den Kunden in der vorgestellten und implementierten Form akzeptiert wird, ist Aufgabe einer noch durchzuführenden Evaluierungsphase. Als „Testumgebung" dient dazu die Arbeitsgruppe Wirtschaftsinformatik. Sollte die Testphase erfolgreich verlaufen, ist der Einsatz der Online-Zeitung für die gesamte Fakultät für Informatik an der Otto-von-Guericke-Universität geplant.

Literatur

[1]   Aßfalg, R.; Zhang, Z. (1997): Implementierung einer WWW-basierten elektronischen Zeitschrift. In: H. Reiterer; T. Mann (Hrsg.): Informationssysteme als Schlüssel zur Unternehmensführung - Anspruch und Wirklichkeit, Konstanz, S. 141-153.

[2]   Benn, W.; Gringer, I. (1996): Datenbank-Anwendungen über das Internet. Informatik-Bericht Nr. 0947-5125, Fakultät für Informatik, TU Chemnitz-Zwickau. Chemnitz.

[3]   Campione, M.; Walrath, K. (1997): Das Java Tutorial: Objektorientierte Programmierung für das Internet. Bonn.

[4]   Dicken, H. (1997): JDBC. Internet-Datenbankanbindung mit JAVA. Bonn.

[5]   Die Zeit (1997): Die Zeit. http://www.zeit.de. 10.12.1997.

[6]   Focus Online (1997): FOCUS Online. http://www.focus.de. 10.12.1997.

[7]   Götze, R. (1994): Dialogmodellierung für multimediale Benutzerschnittstellen. Oldenburg.

[8]   Klute, R. (1996): Das World Wide Web: Web-Server und Clients, HTML 2.0/3.0, HTTP. Bonn.

[9]   Kuhlen, R. (1996): Informationsmarkt: Chancen und Risiken der Kommerzialisierung von Wissen. 2. Aufl., Konstanz.

[10]   Papaj, R.; Burleson, D. (1997): Oracle Databases on the Web. Scottsdale.

[11]   Redmond-Pile, D.; Moore, A. (1995): Graphical user interface design and evaluation (guide): a practical process. Hertfordshire.

[12]   Review of Information Science (RIS) (1997): Review of Information Science (RIS). http://www.inf-wiss.uni-konstanz.de/RIS/. 11.12.1997.

[13]   Stahlknecht, P.; Hasenkamp, U. (1996): Einführung in die Wirtschaftsinformatik. 8. Aufl., Berlin.

[14]   Wirtschaftsinformatik (1997): Zeitschrift Wirtschaftsinformatik. http://www.wirtschaftsinformatik.de/. 11.12.1997.

Adressen der Autoren

Mag. Klement J. Fellner
Otto-von-Guericke-Universität Magdeburg
Fakultät für Informatik
Institut für Technische und Betriebliche
Informationssysteme
Arbeitsgruppe Wirtschaftsinformatik
PSF 4120
39016 Magdeburg

Email: fellner@iti.cs.uni-magdeburg.de
WWW: http://www-wi.cs.uni-magdeburg.de

Dipl.-Kff. Susanne Patig
Otto-von-Guericke-Universität Magdeburg
Fakultät für Informatik
Institut für Technische und Betriebliche
Informationssysteme
Arbeitsgruppe Wirtschaftsinformatik
PSF 4120
39016 Magdeburg

Email: patig@iti.cs.uni-magdeburg.de

Prof. Dr. habil. Claus Rautenstrauch
Otto-von-Guericke-Universität Magdeburg
Fakultät für Informatik
Institut für Technische und Betriebliche
Informationssysteme
Arbeitsgruppe Wirtschaftsinformatik
PSF 4120
39016 Magdeburg

Email: rauten@iti.cs.uni-magdeburg.de

Die Konvergenz von Telekooperationssystemen im Web: Nutzenpotential und Anwendungsfelder

Oliver Reiss

Institut für Wirtschaftsinformatik, Fachbereich Wirtschaftswissenschaften, Philipps-Universität Marburg

Zusammenfassung

Die Konvergenz von Telekooperationssystemen im Internet eine neue Generation von Systemen der computerunterstützten Gruppenarbeit in Aussicht. Neben offensichtlichen technischen Triebkräften dieser Entwicklung sind funktionale und organisatorische Aspekte hervorzuheben, die sich aus der Forderung nach einer besseren Unterstützung der Gruppenarbeit in einer verteilten, rekonfigurierbaren Organisation ergeben.

Gegenwärtig sind die Anfänge einer Entwicklung erkennbar, die im Web Kooperationssysteme hervorbringt, die räumliche Bezugspunkte als virtuelle Orte der Zusammenarbeit verwenden. Diese werden als Entwicklungsziel beschrieben und in möglichen Anwendungsfeldern skizziert.

1 Internet-Technologien als Basis von Kooperationssystemen

Die Forschung im Bereich computerunterstützter Gruppenarbeit (engl. CSCW) war in den vergangenen Jahren in vielen einzelnen Bereichen der Entwicklung und Implementation von Gruppenarbeitssystemen erfolgreich. Trotz zahlreicher Erfolge konnten viele gute Ansätze das Stadium eines experimentellen Prototypen nicht verlassen: Eingeschränkter Funktionsumfang, schwache Integration verschiedener Medien, unzureichende Schnittstellen zwischen Groupware und konventionellen Anwendungssystemen oder eine unzulängliche Anpaßbarkeit der Systeme an praktische Belange sind nur einige Beispiele der immer noch bestehenden Probleme. So brachte das Bemühen um kooperative Anwendungssysteme im wesentlichen zwei Klassen von Lösungen hervor. Zum einen waren dies eigenständige Groupware-Produkte, die auf einer lokalen Vernetzung aufbauend grundlegende, textbasierte Kommunikationsmöglichkeiten boten und in ihrer Nutzung lange Zeit auf asynchrone Nachrichtenfunktionen beschränkt blieben (z.B. Lotus Notes, Novell Groupwise etc.). Zum anderen war die Entwicklung hochspezialisierter Lösungen individueller Gruppenprobleme zu beobachten, die besonderen Erfordernissen einer Arbeitsgruppe in deren individueller Umgebung und Kultur entsprach (z.B. Computergestützte Sitzungs- und Konferenzsysteme) [4].

Internet-Technologien und -standards haben seit ihrem Bestehen einen starken integrativen Einfluß auf die Entwicklung leistungsfähiger kooperativer Computersysteme gehabt. Nicht zuletzt basiert die gegenwärtige Popularität des Internets in der Öffentlichkeit und den Medien auf dessen Eignung, menschliche Beziehungen herzustellen und zu pflegen. Erst seit wenigen Jahren wird jedoch eine Perspektive des produktiven Zusammenwirkens einzelner computerunterstützter Komponenten der Gruppenarbeit im WWW deutlich. Die Perspektive, nicht nur gut strukturierte arbeitsteilige Prozesse durch Internet-Technologien zu unterstützen, sondern auch situativ unterschiedliche Formen spontaner Problemlösungen zu ermöglichen, ist mit der Verfügbarkeit des Webs als standardisierte, offene Plattform und gemeinsamer Bezugsrahmen in greifbare Nähe gerückt. Dabei ist festzustellen, daß sich die gegenwärtige Annäherung von CSCW-Systemen nicht nur in ihren technischen Grundzügen vollzieht, sondern auch in ihren Funktionsprinzipien vor dem Hintergrund ähnlicher Nutzungsformen angleicht.

Die wesentlichen technischen und funktionalen Leitlinien dieser Konvergenz aus Sicht der CSCW-Forschung aufzuzeigen sowie daraus entstehende Anwendungs- und Nutzenpotentiale zu skizzieren, ist Ziel dieses Beitrags.

2 Konvergenztendenzen kooperativer Systeme

Der Begriff der Konvergenz wird in seiner generellen Bedeutung mit der Annäherung und Entwicklung übereinstimmender Merkmale beschrieben. Dabei ist insbesondere der Vorgang der Ausbildung gemeinsamer Wesensmerkmale und das Endergebnis der Übereinstimmung zentraler Strukturelemente kennzeichnend für den Konvergenzbegriff. Das Ergebnis von Konvergenzprozessen ist die Integration der beteiligten Komponenten zu einem übergeordneten Konstrukt. Als Beschreibung sowohl einer sich vollziehenden wie auch einer abgeschlossenen Entwicklung hat der Ausdruck in allen Wissenschaften mindestens eine eigenständige fachliche Bedeutung erlangt.

Im Rahmen dieses Beitrags soll Konvergenz als Vorstufe der funktionalen Integration und der formellen Standardisierung neuartiger Dienste beschrieben werden. Prozesse der Konvergenz sind daher Vorgänge, in denen sich Systeme, Funktionen, Komponenten oder Methoden annähern, ergänzen oder ersetzen. Eine schärfere Definition, die sich meßbarer Quantifizierungskriterien bedient, ist im Lichte der sich noch entwickelnden Strukturmerkmale wenig sinnvoll.

Seit kurzer Zeit wird der Begriff der Konvergenz auch zur Beschreibung der Annäherung von Medien, Technologien und Diensten der Nachrichtentechnik, der Unterhaltungselektronik sowie der Informations- und Kommunikationssysteme verwendet. Eine einheitliche Definition existiert in diesem Zusammenhang (noch) nicht, vielmehr bezeichnet der Begriff in dieser Verwendung eine Reihe von Entwicklungstendenzen, die sich im wesentlichen den folgenden drei Ebenen zuordnen lassen:

- Die *Konvergenz von Basisdiensten der Signalerzeugung, -weiterleitung und -verarbeitung* bezeichnet die Annäherung von Verfahren zur Generierung und dem physischen Transport des (breitbandigen) digitalen Signalstroms. Ausgangspunkte dieser Entwicklung sind zum einen Anforderungen, die sich aus dem Einsatz und der Integration neuer Medien, insbesondere Videoströme, ergeben, und zum anderen die Möglichkeiten, die neuartige Glasfaserübertragungsverfahren erwarten lassen [29].

- Als *Konvergenz von Diensten und Netzzugängen in integrierenden Endgeräten* können die Versuche beschrieben werden, Kommunikationsdienste und Unterhaltungsmedien benutzerfreundlich zusammenzufassen. Erste am Markt verfügbare Geräte dieser Art vereinen Personal Computer inklusive ISDN- und Internet-Zugang mit Geräten der Unterhaltungselektronik wie Fernseher und Stereoanlage.

- Die *Konvergenz von Branchen* ist eine Entwicklung, die auf Basis der Kombination digitaler Medien, Nachrichten- und Kommunikationsdienste neuartige Produkte und Dienstleistungen hervorbringt. Gegenwärtig ist die Entstehung von Anwendungsfeldern integrierter Informationsquellen aus z.B. Telekommunikation, digitalem Rundfunk und Informationsverarbeitung zu beobachten, die auch als *Mediamatik* bezeichnet werden [17].

Konvergenzen finden jedoch nicht nur in den oben beschriebenen Ebenen statt, sondern auch innerhalb der jeweiligen Netze und Dienste. Besonders das WWW als bedeutsamster Internet-Dienst zeigt das Zusammenwachsen von wesentlichen Einzeldiensten und -funktionen an zahlreichen Stellen auf. In kaum einem anderen Bereich des Webs wird dieses zur Zeit so deutlich wie bei der Entwicklung leistungsfähiger kooperativer Anwendungen. Die Verfügbarkeit einer gemeinsamen, offenen Plattform für Telekooperationssysteme zur "medien-gestützten, arbeitsteiligen Leistungserstellung von individuellen Aufgabenträgern, Organisationseinheiten und Organisationen, die über mehrere Standorte verteilt sind " [23, S. 52] verspricht Organisationen zahlreiche Möglichkeiten der Überwindung zeitlicher, räumlicher und institutioneller Barrieren. Gegenwärtig sind die Anfänge einer Entwicklung zu erkennen, die in zunehmendem Maße kooperative Anwendungen wie Workflow-Management-Systeme, Konferenzsysteme, Media Spaces, Telepräsenzsysteme, synchrone und asynchrone Kommunikationssysteme in das Web portiert, sofern sie nicht bereits auf dessen Basis entwickelt wurden und deren Annäherung aneinander in bezug auf unterstützte Standards, vergleichbare Funktionen und Kooperationsformen bereits mehr oder weniger deutlich wird. Stand zu Beginn dieser Entwicklung noch die Abbildung von klassischen Groupware-Funktionen im Web im Vordergrund, entstehen heute zunehmend neue Anwendungssysteme, die nicht nur *eine* bestimmte Kooperationsform unterstützen, sondern auch weitere zentrale Anforderung an CSCW-Systeme erfüllen: die Unterstützung eines breiteren Spektrums von Gruppenarbeitssituationen sowie die Unterstützung der Übergänge zwischen diesen Situationen [28, S. 233]. Als konkretes Beispiel ist die Ergänzung von expliziten Koordinationssystemen (Workflow Management Systeme) durch implizite Kooperationssysteme (Desktop Video Conferencing Systeme) zu nennen, wie sie bereits in Prototypen und proprietären Anwendungssystemen vollzogen wurde [27].

2.1 Technische Einflußfaktoren von Internet-Technologien auf das Zusammenwachsen von Telekooperationssystemen

Zentrale Antriebskräfte der Konvergenz von Telekooperationssystemen sind zunächst in dem wachsenden Reifegrad von Internet-Basistechnologien zu sehen. Neben einer ständigen Verbesserung der IP-basierten Übertragungsverfahren in der Abstimmung mit Komponenten der physischen Netzwerkschicht ist insbesondere die zunehmende Verfügbarkeit von leistungsfähigen Verfahren zur Steuerung und Administration von IP-Signalströmen zu bemerken.

Zunächst sollen einige grundlegende Einflüsse von Internet-Technologien genannt werden, die nicht spezifisch für Telekooperationsanwendungen sind, aber zentrale Bausteine bei der Konzeption dieser Systeme darstellen.

Grundlegende Einflüsse von Internet-Technologien (Konvergenz der Basistechnologien)

Als wichtigsten technischen Einflußfaktor ist zunächst die umfassende zügige Standardisierung von Schlüsseltechnologien und -verfahren zu sehen. Dieses betrifft dabei sowohl Basistechnologien als auch Spezifikationen wie Dokumentenstruktur, Verzeichnisdienste, Nachrichtendienste oder Audio-/Videokommunikation. Die Zahl verwendbarer Protokolle und Standards zur Übertragung und Verarbeitung von Informationen deckt dabei immer mehr Basisfunktionen ab, die bis dato nur aus proprietären Groupware-Systemen bekannt waren. Die Erläuterung von Details und Leistungsmerkmalen von Internet-Standards wie HTML4, IMAP4, LDAP, UDL oder iCAL erscheint vor dem zeitlichen Hintergrund der mittel- bis langfristig ablaufenden Konvergenz an dieser Stelle nicht sinnvoll.

Eine entscheidende Rolle bei der Konvergenz von Telekooperationssystemen im Web spielt die Entwicklung von leistungsfähigen verteilten Objektsystemen. Komplexe Anwendungen und Funktionen der Gruppenarbeit erfordern einen Grad der Anwendungsentwicklung, der gegenwärtig von dem in weiten Teilen durch CGI, proprietäre Datenbank-Zugänge und Applet-Mechanismen geprägten Web nur eingeschränkt gewährleistet werden kann. Elementare Anforderungen an Informations-Objekte wie Konsistenz, Persistenz oder Grundfunktionen des Datenbank-Managements zwischen Anwendungsklassen verlangen nach einer Client-Server-Architektur, deren Konturen sich bereits zu formen begonnen haben. Derartig grundlegende und erfolgversprechende Entwürfe liegen zur Zeit in Form der OMG Common Object Request Broker Architecture (CORBA), SUNs Java/JavaBeans und Microsofts Active X/DCOM vor. Mittelfristig kann CORBA dabei die Funktion einer offenen Objektinfrastruktur für 'mobile' Codesysteme wie Java zukommen, die die Gestaltung modularer Funktionen erlaubt, die heute lediglich aus leistungsfähigen Middleware-/Transaction Processing Systemen bekannt sind [21, S. 50]. Deren Existenz und Funktionsweise wird auch den für kooperative Anwendungssysteme wichtigen Dokumentenbegriff entscheidend beeinflussen, da Ansätze wie der *Compound Document Container* auf einem wohldefinierten Objektbegriff und die zugehörigen Funktionen angewiesen sind [20, S. 270].

Spezifische Einflüsse von Internet-Technologien auf kooperative Anwendungen (funktionale Konvergenz)

Die Annäherung unterschiedlicher Kooperationssysteme wird durch die Verfügbarkeit gemeinsamer Standards der Abwicklung von Kommunikationsprozessen drastisch beschleunigt. Kommunikations- und Konferenzsysteme können auf einer großen Zahl bewährter synchroner und asynchroner Kooperationsformen auf der Basis von Internet-Technologien aufbauen. Insbesondere Notifikations- und Informationsvisualisierungsstandards auf Basis des Hypertext-konzeptes stellen bereits eine gute Grundlage des Zusammenwirkens kooperativer Anwendungen dar. Die Möglichkeiten, über diese Kanäle unstrukturierte Kommunikation zu betreiben sowie über verschiedene Medien kontextuelle Informationen der Bewältigung der Gruppenaufgabe auszutauschen, sind vergleichsweise einfach zu implementierende Funktionen, die in Kürze zur Grundausstattung kooperativer Anwendungen im Web gehören dürften. Dabei werden eine Reihe dieser Kommunikationswerkzeuge bereits über den Browser verfügbar sein und somit nur noch von Telekooperationsanwendungen gesteuert werden müssen.

Die Möglichkeit, über Plattform- und Funktionsgrenzen Objekte auszutauschen und im Mehrbenutzerbetrieb Synchronisations- und Sicherheitsregeln einzuhalten, dürfte die Schlüsselfunktion der Gestaltung sowohl intelligenter Nachrichtensysteme als auch komplexer, verteilter Vorgangssteuerungssysteme sein. Das. Zusammenwirken von unterschiedlichen Kooperationssystemen, die beispielsweise Informationen über Vorgangszustände auf Instanzebene, Zugriffsregeln oder komplette geteilte Objekte austauschen, ist dabei von der Funktionsfähigkeit gegenwärtig entstehender offener Verzeichnisdienste, verteilter Objektmanagementsysteme und deren Datenbanken abhängig [1]. Eine enge Integration unterschiedlicher Telekooperationsanwendungen, wie etwa Workflow Management Systeme, Gruppeneditoren und Desktop Conferencing Systeme ist somit direkt an einen breiten Funktionsumfang eines leistungsfähigen offenen, verteilten Objektsystems gebunden, das nicht nur eine Plattform für den Austausch von gemeinsamen Steuerflüssen ist, sondern vielmehr auch einzelne Artefakte oder ganze Kontextobjekte der Kooperation zu verwalten vermag.

Die sich abzeichnenden Architekturen von Software-Komponenten im Web zeigen eine weitere Konvergenzlinie auf. Durch die Bündelung einzelner Funktionen des Objekt-Managements zu Komponenten innerhalb einer vorgegebenen Architektur kann der Systementwicklungsprozeß kooperativer Anwendungen beschleunigt und durch Wiederverwendung kostengünstiger gestaltet werden. Wird ein allgemeines Architekturmodell eines kommunikationsorientierten Gruppenarbeitssystems betrachtet [32, S. 97 ff.], so wird offensichtlich, daß sich die Grundfunktionen des Systemkerns auch durch wiederverwendbare Softwarekomponenten gestalten lassen können. Komponenten wie Aufgabenkoordination, Kontextverwaltung, externe/interne Informationsverwaltung, Kommunikationssteuerung und Steuerung externer Applikationen können als Gesamt- oder Teilkomponenten in ein kooperatives Anwendungssystem übernommen werden. Die sich abzeichnende Entwicklung von Komponenten-basierten Architekturen [15] im Web dürfte daher Funktionsmodule kooperativer Anwendungen hervorbringen, die auf der Basis einheitlicher Datentypen und Objekt-Managementregeln das Zusammenwachsen von Systemen der computerunterstützten Gruppenarbeit fördern.

2.2 Begründung und Antriebskräfte der Konvergenz kooperativer Systeme aus Sicht der Unternehmensorganisation

Neben technischen Triebkräften, die die Konvergenz von Telekooperationssystemen im Web begünstigen, sind es insbesondere Bedürfnisse moderner, verteilter Organisationen, die zur Weiterentwicklung leistungsfähiger Kooperationssysteme in diesem Sinne beitragen dürften.

Hierarchische Organisationsstrukturen haben sich unter dem Wettbewerbsdruck dynamischer, globaler Märkte als zu unflexibel erwiesen. Um einfachere Wege der Leistungserstellung zu finden, Wertschöpfungsketten mehrerer Partnerorganisationen miteinander zu verbinden und flexibler auf Kunden- und Markterfordernisse reagieren zu können, entstehen modulare und virtuelle Organisationsstrukturen, deren Aufgabenintegration per Gruppenarbeit erfolgt [24, S. 226]. Dieser kurzfristige Aufbau von zeitlich befristeten Organisationseinheiten soll ein Maximum an Flexibilität und Reaktionsfähigkeit bewirken und die kurzfristige Rekonfigurierbarkeit von Organisationsstrukturen gewährleisten. Eine möglichst flexible

Umgestaltung ablauf- *und* aufbauorganisatorischer Strukturen muß dabei vor dem Hintergrund genau festgelegter Kernbereiche der Geschäftstätigkeit sowie deren zentralen Wertschöpfungsprozessen erfolgen, so daß die Gefahr von Reibungsverlusten und Ressourcenverschwendung vermieden wird [19, S. 3]. Eine eindeutige Abgrenzung der durch das Unternehmen selbst zu erbringenden Leistungen stärkt den Wert von zwischenbetrieblichen Kooperationen. So kommt der Nutzung von Partnerschaften zur Sicherung von Kapazitäten, die die Organisation nicht besitzt, besondere Bedeutung zu [9].

Vor diesem Hintergrund werden integrierte Werkzeuge und kooperative Anwendungssysteme bedeutsam, die einen möglichst breiten Raum zur individuellen Problemlösung der Gruppe einräumen, mächtige Kommunikationsmedien zur Verfügung stellen und der Gruppe die Verwendung eigener Koordinationsmechanismen zur Selbstorganisation erlauben. Des weiteren sind Kooperationssysteme erforderlich, die den Mitgliedern ein Höchstmaß an Transparenz der um ein Vielfaches komplexeren Organisationsstruktur einräumen. Neben dem daraus resultierenden Bedarf einer leistungsfähigen Kommunikationsinfrastruktur entsteht das Erfordernis, den organisatorischen Kontext der Gruppenarbeit so umfassend wie möglich abzubilden [26, S. 120] [22, S. 379]. Dieses bedeutet, daß es für jeden Prozeßbeteiligten möglich sein muß, zu jedem Zeitpunkt und an jedem Ort der Leistungserstellung die für seine Arbeit erforderlichen Kontextinformationen zu erlangen und Sinn und Inhalt seiner Aufgaben innerhalb der Gruppe zu verstehen [18]. Des weiteren birgt die räumliche und zeitliche Loslösung arbeitsteiliger Prozesse ein radikales Veränderungspotential der sozialen Struktur der Organisation in sich, der durch die umfassende Gestaltung des formellen und informellen Kontextes der Arbeit Rechnung getragen werden muß [25].

Die hier umrissenen Entwicklungen organisatorischen Wandels begründen umfangreiche Anforderungen an verteilte Kooperationssysteme, die Hilfsmittel für verschiedene Kooperationssituationen, integrierte Objekttypen und -funktionen mit einem unkomplizierten Zugang verbinden.

2.3 Verteilte virtuelle Räume in Netzgemeinschaften als mögliches Entwicklungsziel

Nicht zuletzt als Reflex auf starre Vorgangssysteme, die lediglich Prozesse automatisieren, den Kontext der Zusammenarbeit aber vernachlässigen [2], sind in den vergangen Jahren eine Reihe von Kooperationssystemen aus ambitionierten Forschungsprojekten hervorgegangen, die neben der Verbindung leistungsfähiger Werkzeuge der Audio- und Videokommunikation auch die Wahrnehmungsinformationen (*awareness*) von Vorgängen im Umfeld der aktuellen Kooperation unterstützten, z.B. [13, S. 84 ff.]. Die Idee von *media spaces*, bzw. deren Erweiterung *collaborative virtual environments,* basiert auf der Vorstellung, einer Gruppe mindestens einen gemeinsamen elektronischen, nach Möglichkeit dreidimensionalen Kooperationsraum zu geben. Dessen Konzeption und Realisierung brachte in der CSCW eine große Zahl von Ansätzen und Prototypen hervor, die sich grob in die Kategorien *meeting points, social places* und *working spaces* unterteilen lassen [11]. Sie stellen eine Forschungsrichtung dar, die aufgrund der Verwendung von gemeinsamen Räumen (*space*) und Orten (*place*) als Basis der Interaktion von Kooperationspartnern eine starke integrative

Wirkung auf spezifische Gruppenunterstützungssysteme ausübt. Während der Raum (space) derjenige Begegnungsbereich ist, in dem Interaktionen überhaupt möglich werden, spiegelt der Ort (place) die konkrete Vorstellung der Gruppe wieder, auf welche Weise die Zusammenarbeit erfolgen soll. [12] Die Verwendung von physischen Metaphern wie dem Raum geben Kooperationswerkzeugen eine Klammer, sie binden Kontext und Artefakte der Zusammenarbeit in einem Medium. Räume wirken somit wie Container, die Akteure und Objekte der Zusammenarbeit enthalten.

Die Nutzung eines gemeinsamen Interaktionsraumes bietet eine Reihe von Vorteilen, die sich hauptsächlich aus dem Vorhandensein eines gemeinsamen Bezugspunktes der Kooperation ergeben, in dem verschiedene Elemente des Gruppenarbeitskontextes dargestellt und integriert werden können. Das Raum-Konzept ist dabei geeignet, als Brücke über die klassischen Gräben kooperativer Anwendungssysteme zu dienen. Dieses sind z.B. Lücken zwischen

- individueller und gemeinsamer Arbeit,
- synchroner und asynchroner Kommunikation,
- einzelnen Phasen des Arbeitsprozesses,
- räumlichen Entfernungen einzelner Gruppenmitglieder,
- Groupwarefunktionen (z.B. joint editing) und individuellen Anwendungswerkzeugen (z.B. Textverarbeitung) [11].

Wünschenswert wäre somit die Zusammenfassung verschiedener Interaktionsinformationen und -hilfsmittel wie der Transparenz der Organisationsstrukturen, das Angebot verschiedener Kooperationsformen zur individuellen Problemlösung sowie die Bereitstellung von Arbeitsmitteln und -werkzeugen zur Durchführung des eigenen Arbeitsprozesses und der Wahrnehmung anderer Arbeitsprozesse in einer Organisation [8, S. 15].

Eine solche Umgebung bietet in einer adäquaten visuellen Umsetzung des weiteren den Vorteil der Nutzung der menschlichen Affinität für grafische Informationen. Neben einem spielerischen Umgang fällt dem Benutzer insbesondere die Navigation und die Wiedererkennung wichtiger Elemente leichter. Die Erfolgsgeschichte der Desktop-Metapher als Grundlage von Fenster-Systemen läßt vermuten, daß eine Schnittstellengestaltung, die noch mehr Möglichkeiten der Modellierung vertrauter Umgebungsinformationen erlaubt, auch vom Benutzer angenommen werden dürfte. Wie jedoch Mißerfolge erster Prototypen und Produkte zeigen, kann es keineswegs das Ziel sein, eine exakte Entsprechung von realen Kooperationsstrukturen oder Umweltbedingungen zu modellieren, sondern lediglich vertraute Analogien der Zusammenarbeit zu schaffen [14, S. 58 ff.]. So ist es beispielsweise wünschenswert, den Zugang zu Räumen, in denen vertrauliche Informationen ausgetauscht werden, mit einem Türmechanismus zu beschränken, während physische Komponenten wie Korridore als Verbindung zwischen Räumen überflüssig werden.

Im Internet haben sich seit dessen Entstehung bereits gemeinsame Foren und Plätze der Kooperation gebildet, die bis zur Entwicklung des Webs auf den einfachen textbasierten gleichzeitigen oder zeitversetzten Nachrichtenaustausch beschränkt waren und auf gemeinsame Hilfsmittel und geteilte Artefakte der Zusammenarbeit verzichten mußten. Kommerzielle On-line-Dienste und das WWW brachten in den letzten fünf Jahren eine Reihe unterschiedlicher Kommunikationsforen hervor, die die ersten Anfänge einfacher Bulletin

Board-, Relay Chat-, MUD- und MOO-Systeme schnell hinter sich ließen. Während mit anfänglichen Informationsdiensten wie *The Well, ECHO, Parent Soup* etc. die Portierung textbasierter Interaktion in einen Web-Client vollzogen wurde, entstanden innerhalb der letzten Jahre vermehrt Kommunikationsforen, die räumliche Umgebungsabbilder zum Fokus der Interaktion zwischen Teilnehmern machten. *The Palace,* eines der ersten Systeme dieser Art, bietet eine Reihe von primitiven Landschaften, in denen Benutzer, durch ein mehr oder weniger aussagekräftiges Symbol gekennzeichnet (*avatar*), einfache Kommunikationshandlungen durchführen können. So einfältig diese Form der Interaktion auch für arbeitsteilige Vorgänge erscheinen mag, ist erkennbar, daß bereits dieses simple Raumsystem eine Plattform für Koordinationsprozesse zwischen beteiligten Akteuren sein könnte.

Einen mächtigeren und erfolgversprechenderen Ansatz der Gestaltung interaktiver räumlicher Umgebungen stellt die Virtual Reality Modelling Language dar, die seit Ende letzten Jahres in dem ISO-Standard *VRML 97* vorliegt. Weniger eine Programmiersprache wie *Java3D* denn mehr ein Beschreibungsformat, erlaubt VRML die Modellierung dreidimensionaler Strukturen. Die ursprüngliche Vorstellung, VRML innerhalb kurzer Zeit zu einer Entwicklungsplattform weiterzuentwickeln, die nach der rudimentären dreidimensionalen Informationsvisualisierung auch die Implementation von Mehrbenutzerinteraktionen sowie der Integration externer Funktionalitäten wie audio-visuelle Datenströme, Java-Applikationen, relationaler Datenbanken etc. in einer einheitlichen Spezifikation vereint, konnte nicht realisiert werden [16, S. 290 f.]. Vielmehr zerfielen die gemeinsamen Bemühungen des Konsortiums in unterschiedliche Einzelrichtungen, die eine Reihe proprietärer Produkte, wie z.B. *Living Worlds* hervorbrachten. Dennoch ist es gelungen, einen einheitlichen Standard der Beschreibung strukturierter Mengen dreidimensionaler Bilder (Szenen) zu schaffen. [16, S. 341] bezeichnen diesen als das "Postscript des Cyberspace". Auch die Implementation von Multiuser-Fähigkeiten in VRML ist mittlerweile derartig vorangeschritten, daß sich dieser Standard als Grundlage zukünftiger Interaktions-Systeme etablieren könnte [3]. Die auf der Basis von VRML und deren Erweiterungen wie denen von *Sony, blaxxun* oder *Worlds* entwickelten virtuellen Gemeinschaften stellen Ansätze zur Gestaltung derjenigen ‚sozialen Klammern' dar, die die CSCW mit ihren Forschungen in den Bereichen *media spaces* und *collaborative virtual environments* vorgegeben hat.

Seit letztem Jahr sind die ersten Web-basierten Groupware-Systeme am Markt verfügbar, die auf der Verwendung eines gemeinsamen Informationsraumes basieren. Beispielhaft seien *Netopia Virtual Office, Changepoints involv Intranet, HotOffice* oder *Instictive eRoom* genannt. Diese Anwendungen basieren darauf, daß sich Kooperationsteilnehmer eigene Räume, Foren oder Arenen einrichten, in denen sie Objekte der Bürokommunikation sowie Nachrichten austauschen können. Neben Grundfunktionen wie Dokumentenmanagement-Möglichkeiten werden verschiedene Internet-Messaging-Dienste und Schnittstellen zu Konferenz-Diensten wie z.B. UDL-basierter Videokommunikation innerhalb der Räume angeboten. Dabei variiert die technische Lösung zwischen einer Erweiterung der Browser-Funktionalität um Server-Dienste und der kompletten Abwicklung des Raum-Managements auf dem Server eines Service-Providers, der Interaktionsräume *vermietet.* Die gegenwärtig vorherrschende Verwendung einer zweidimensionalen Raum-Metapher könnte, wie bereits in ersten Prototypen eingeschränkter Interaktion deutlich, dem dreidimensionalen VRML-basierten Interaktionsraum weichen. Daß dazu neben einem leistungsfähigeren Client-/Server-Modell im Web auch eine große Zahl von Schnittstellen- und Bandbreitenproblemen zu lösen ist, ist selbstverständlich. Entscheidend ist jedoch, daß sich die Verwendung einer *Klammer* zur Integration von Gruppenkommunikation und Kooperationskontext, wie sie bereits aus CSCW-Forschungen bekannt ist, in diesen Systemen im Web zeigt.

Neben der Schaffung völlig neuer Interaktionsformen in Kooperationsräumen im Web stellt auch der Wunsch nach einer besseren Abbildung realer Verhältnisse und deren Funktionen auf entsprechende virtuelle Umgebungen eine Begründung raumbasierter Anwendungen dar. Insbesondere Unternehmen, die ein einheitliches Auftreten auf realen und elektronischen Märkten anstreben, suchen nach Wegen, physisch vorhandene Organisationsstrukturen im Web abzubilden. Das Beispiel eines elektronischen Firmenfoyers im Web als Spiegel eines realen Eingangsbereiches verdeutlicht Anforderungen an Interaktionsfähigkeit, Informationsvisualisierung und Sicherheitsmechanismen, die grundsätzlich von virtuellen Räumen erfüllt werden können [2, S. 82 f.].

Daß sich als konkrete Form dieser Klammer dreidimensionale Kooperationsräume auf Basis des Object Web zumindest in Prototypen zeigen werden, erscheint sehr wahrscheinlich, sofern es gelingt, die aktuellen Ansätze der Beschreibung gemeinsamer Umgebungen im Sinne dreidimensionalen Welten zu einer mit anderen Internet-Schlüsseldiensten integrierten Entwicklungsplattform auszubauen.

3 Verteilte virtuelle Räume: Ausgewählte Nutzenaspekte anhand von Beispielen aus dem Dienstleistungssektor

Die sich abzeichnende Konvergenz Web-basierter Telekooperationssysteme soll im folgenden anhand von Beispielen skizziert werden. Dabei steht die Betrachtung des Nutzens derartiger Systeme für eine modulare verteilte Netzwerk-Organisation im Vordergrund [24, S. 226 ff.]. Es werden Anwendungsmöglichkeiten innerhalb von Dienstleistungsunternehmen beschrieben, deren Leistungserbringung mit hohem Koordinations- und Kooperationsaufwand sowie Spezialisierungsgrad verbunden ist [5] und in denen gemeinsam genutzte digitalisierte Informationen Arbeitsformen begründen. Ohne an dieser Stelle zwingend ein Integrationsmodell aufzustellen, können in Anlehnung an [1] die folgenden Ebenen der Kooperation unterschieden werden:

- *Basiskooperationen am Arbeitsplatz,*
- *Intraorganisatorische Kooperationsgemeinschaften,*
- *Extra- und interorganisatorische Kooperationsgemeinschaften.*

3.1 Restrukturierung und Repräsentation kooperativer Prozesse am Arbeitsplatz eines Ingenieurbüros

Auf der Ebene der *Basiskooperation am Arbeitsplatz* kann die Verwendung von an die Realität angelehnten Repräsentationsformen die Benutzbarkeit des kooperativen Systems sowie die Navigation innerhalb von Aufbauorganisations- und Datenstrukturen deutlich steigern. So könnte der Arbeitsplatz eines Ingenieurs, dessen Zugehörigkeit zu Projektgruppen stets befristet ist, mit Hilfe von VRML-basierten Visualisierungsmethoden als persönlicher

Informationsraum gestaltet werden, der einen zentralen Zugang zu Werkzeugen der fachlichen Produktivität (CAD-Systeme), der Bürokommunikation und individuellen Sichten auf verteilte Informationsbestände der Organisation verwirklicht. Dieser persönliche *home room* dient als Bindeglied zwischen einzelnen Kooperationssystemen, die über die Einbettung von Internet-Kommunikationsstandards und Java-Applikationen bereits heute realisiert werden können. Wie bereits aus zweidimensionalen CSCW-Raumsystemen bekannt, werden an einem bestimmten Platz des Raums dynamisch Umgebungsinformationen dargestellt, die die erforderliche Wahrnehmung informeller, gruppen-struktureller, sozialer und vorgangsspezifischer Informationen gewährleistet [10]. Aus dem persönlichen home room führen festgelegte organisationsspezifische Wege zu Informationsbeständen, die den Kontext der individuellen Tätigkeit beschreiben. Beispielhaft sei der dreidimensionale Zugang zu einem Organisationsdiagramm der gegenwärtigen Leistungserstellung in einem Raum oder dem Wissensbestand einer Organisation in einem anderen Raum erwähnt. Die Verwendung der Raum-Metapher erlaubt dabei auch die funktionale Erweiterung von Kooperationsfunktionen: So weicht eine Nachrichten-Eingangsbox einem persönlichen Vorraum, in dem Kooperationsmitglieder Artefakte der Zusammenarbeit ablegen, aber auch die gegenwärtige *Ansprechbarkeit* mit dem Ziel des Aufbaus einer ad-hoc-Konferenz überprüfen können.

Zur Überwindung räumlicher Restriktionen sowie zur Erweiterung des Sicherheits-Spielraumes ist es denkbar, einen derartigen persönlichen Informationsraum mit Hilfe einer Chipkarte portabel zu gestalten. Der Benutzer ist somit nicht an festgelegte Orte und Werkzeuge der Leistungserstellung im Sinne eines konventionellen Arbeitsplatzes gebunden. Ein weiterer Vorteil ist in den erweiterten Möglichkeiten einer Organisation zu sehen, Transparenz in die gegenwärtige Organisation und aktuell ablaufende Leistungserstellung innerhalb von virtuellen Einheiten zu bringen, da die Chipkarte als aktive Komponente auch Auskunft über Aktivitäten des Benutzers im allgemeinen und über Vorgangszustände im besonderen geben kann.

Erste Ansätze und Prototypen Web-basierter persönlicher Kooperationsräume orientieren sich dabei zumeist noch an physischen Gegebenheiten eines Büros (Schreibtisch, Aktenablage etc.). Jedoch zeigen die im WWW entstehenden raumbasierten Groupware-Applikationen, daß das Festhalten an konventionellen realen Verhältnissen im einfachsten Fall nicht notwendig, im schlimmsten Fall sogar hinderlich sein kann [14, S. 61].

3.2 Auflösung physischer Arbeitsplätze und Virtualisierung von Prozeßketten in Projektgruppen der Unternehmensberatung

Ausgangspunkt von Dienstleistungsprozessen in Unternehmensberatungsgesellschaften ist die Zusammensetzung einer Projektgruppe, die nach erfolgreicher Problemlösung wieder auseinandergeht. Die Unterstützung dieser *intraorganisatorischen Kooperationsgemeinschaften* verlangt Werkzeuge, die über ein gemeinsames Dateisystem mit separaten Kommunikationskanälen weit hinausgehen. Raumbasierte Telekooperationssysteme können hier den kurzfristigen, flexiblen Aufbau von gemeinsamen Benutzerforen zur Koordination arbeitsteiliger Prozesse und der Sicherung von im Projektverlauf gewonnen Erkenntnissen ermöglichen. So können sich Projektgruppen über deren gesamten

Lebenszyklus hinweg eigene Kooperationsräume schaffen, in denen Objekte ausgetauscht, Informationen archiviert und ad-hoc Probleme mittels verschiedenster Kommunikationshilfsmittel gelöst werden. Raumbasierte CSCW-Werkzeuge wie *TeamWave* [11] zeigen, daß grundsätzlich alle wichtigen Interaktionsformen wie Person-Person, Gruppe-Person und Gruppe-Gruppe einbezogen werden können, so daß die Prozeßkette der Leistungserstellung in hohem Maße virtualisiert werden kann [31, S. 5]. Prototypen wie das *KMi Stadium* des *Knowledge Media Institutes* [6] zeigen, daß bereits mit heutigen Web-Technologien Interaktionsräume auch für große Personenzahlen konstruiert werden können, so daß der Raumbegriff skaliert werden kann und eine große Bandbreite für die in raumbasierten Systemen ablaufenden Kooperationsprozesse möglich ist.

Die Visualisierung des Projektfortschritts ermöglicht dabei auch die umfassende Einbindung des Kunden in den Projektablauf, der den gesamten Beratungsprozeß auf Wunsch als Besucher des Raumes miterleben und in bezug auf Kosten und Nutzen einen konkreten Einblick in die erledigte Arbeit gewinnen kann.

Eine mögliche technische Implementierung derartiger Kooperationssysteme könnte auf Basis von *shippable places* erfolgen, einem Container für CORBA-Objekte, der Internet-Dienste zur Kommunikation mit der Broker-gestützten Koordination von Business Objekten verbindet [20, S. 284 f.].

3.3 Interorganisatorische Verbindung von Wertschöpfungsketten auf Basis gemieteter virtueller Räume am Beispiel einer Werbeagentur

Extra- und interorganisatorische Kooperationsgemeinschaften beschreiben die Ebenen, in denen Beziehungen zwischen Organisationseinheiten verschiedener Unternehmen eingegangen werden. Dieses kann sich sowohl auf die Einbindung eines einzelnen, meist dauerhaften Kooperationspartners beschränken (*extraorganisatorische Zusammenarbeit*) wie auch die Nutzung von Unternehmensnetzwerken zur zeitlich befristeten Projektarbeit im Sinne virtueller Unternehmen [5] beabsichtigen (*interorganisatorische Zusammenarbeit*).

Ein anschauliches Beispiel für die Anwendung von Kooperationsräumen ist die Durchführung eines Kundenauftrages einer Werbeagentur. Nachdem über einen geeigneten Internet-Broker ein Partner zur Produktion und Verteilung von Werbemedien ausfindig gemacht wurde [7, S. 4], mietet die Projektgruppe der Agentur geeignete Kooperationsräume eines Internet Service Providers, der erforderliche Arbeitsbereiche, technische und organisatorische Sicherheitsmechanismen und eine Infrastruktur zur Steuerung synchroner und asynchroner Kommunikation bereitstellt. In enger Zusammenarbeit mit dem Kooperationspartner werden die erforderlichen Ergebnisse im Kooperationsraum erarbeitet und dieser nach Abschluß des Projektes inklusive der Projekthistorie in eine Bibliothek zur Wiederverwendung übernommen. Wird im Verlauf der Zusammenarbeit deutlich, daß aufgrund guter Aufgabenintegration und positivem Projekt-ausgang eine dauerhafte Zusammenarbeit beider Partner wünschenswert ist, kann in einem der beiden Unternehmen ein langfristig angelegtes Kooperationsforum als System verbundener Räume eingerichtet werden, das eine plattformunabhängige Verbindung auf der Basis der oben beschrieben Web-Technologien beider Seiten ermöglicht.

Nachdem Lotus Mitte letzten Jahres mit *Instant TeamRooms* ein erstes raum-basiertes Kooperationssystem bei verschiedenen Internet Service Providern zur Miete anbot, sind mittlerweile eine Reihe von erheblich leistungsfähigeren Diensten am Markt erhältlich. Diese bieten nicht nur einfache Messaging-Dienste, Diskussionsforen und serverseitige Speicherplätze, sondern auch simple Workflow Management-Funktionen und die Handhabung von Dateiformaten gängiger Endbenutzerwerkzeuge wie Word, Excel etc. mit schlichten Zugriffskontrollmechanismen [30].

4 Schlußbetrachtung

Der vorliegende Beitrag zeigt Entwicklungslinien der Konvergenz von Telekooperationssystemen im Web auf. Dabei weisen heutige Koordinations- und Kooperationssysteme grundsätzlich die gleichen Probleme wie andere Anwendungssysteme im WWW auf. Das Fehlen einer grundlegenden Architektur zur Entwicklung von Client-/Server-Applikationen ist das entscheidende Charakterisierungsmerkmal aktueller Kooperationssysteme im Web, die zwar bereits aus einer großen Zahl verschiedener Kommunikationsdienste und spezifischen Problemlösungen bestehen, die aber weder technisch noch funktional sinnvoll integriert können. Die bereits seit längerer Zeit in den Forschungsarbeiten zu computerunterstützter Gruppenarbeit und Human-Computer-Interaction diskutierten Konzepte von Kooperationsräumen sind zunehmend auch im Web erkennbar. In erster Näherung ist mit dieser Metapher ein Bindeglied zu erkennen, das sich als mächtiges Leitbild der Gestaltung zukünftiger Telekooperations-systeme im Internet abzeichnet. Ob dieses aus Sicht aktueller Entwicklungen im Bereich der Unternehmensorganisation geeignet ist, ein soziales Umfeld und den Kontext der Aufgabenerfüllung virtueller Organisationseinheiten abzubilden, muß die empirische Forschung zeigen.

Literaturverzeichnis

[1]   W. Augsburger u.a.: Integration von CSCW-Anwendungen zur Unterstützung von Telekooperation. In: H. Krcmar, H. Lewe, G. Schwabe (Hrsg.): Herausforderung Telekooperation: Einsatzerfahrungen und Lösungsansätze für ökologische, technische und soziale Fragen unserer Gesellschaft. Berlin u.a. 1996, 243-257

[2]   S. Benford u.a.: Shared Spaces: Transportation, Artificiality, and Spatiality. In: ACM (Hrsg.): CSCW '96: 6th Conference on Computer-Supported Cooperative Work. Cambridge, USA 1996, 77-86

[3]   W. Broll, J. Fecher, D. Schick: Unterstützung von Multiuser-VR mit VRML. In: Proceedings of the first GI Workshop Assistants, Agents, Avatars (AAA'97). Darmstadt 1997, 9-17

[4]   D. Crow, S. Parsowith, B. Wise: Students: The Evolution of CSCW - Past, Present and Future Developments. In: SIGCHI Bulletin Nr. 2, April 1997, 20-26

[5]   W.H. Davidow, M.S. Malone: The Virtual Corporation: Structuring and Revitalizing the Corporation for the 21th Century. New York 1992

[6]   M. Eisenstadt: About KMi Stadium: Web-based Audio-/Visual Interaction as Reusable Organisational Expertise. In: Workshop on Knowledge Media für Improving Organisational Expertise, 1[St] International Conference on Practical Aspects of Knowledge Management, Basel, 30.-31. Oktober 1996

[7]   W. Faisst: Information Technology as an Enabler of Virtual Enterprises: Life-cycle-oriented Description. In: IT Vision'97, The European Conference on Virtual Enterprises and Networked Solutions, CD-Rom. Paderborn 1997. URL: http://orgbrain.wi1.uni-erlangen.de/public/deutsch/veroeffentlichungen/ spezial/vu/fai_abs13.html, Abruf am 97-12-22

[8]   L. Fuchs, W. Prinz: Aspects of Organisational Context in CSCW. In: Schmidt, K., Bannon, L.: Issues of Supporting Organizational Context in CSCW Systems. Lancaster 1993, 11-48

[9]   J. Galbraith: The Reconfigurable Organization. In: Hesselbein, F. u.a. (Eds.): The Organization of the Future. San Francisco 1997, 87-98

[10]   S. Greenberg, C. Gutwin, A. Cockburn: Using distortion-oriented displays to support workspace awareness. In: A. Sasse, R.J. Cunningham, R. Winder (Hrsg.): People and Computers XI (Proceedings of the HCI'96). London u.a. 1996, S. 299-314

[11]   S. Greenberg, M. Roseman: Using a Room Metaphor to Ease Transitions in Groupware. Research report 98/611/02, Department of Computer Science, University of Calgary, Calgary, Alberta, Canada, January. URL: http://www.cpsc.ucalgary.ca/projects/grouplab/papers/98-RoomMetaphor/ report_98_611_02/ room_metaphor.html, Abruf am 98-03-01

[12]   S. Harrison, P. Dourish: Re-place-ing space: the roles of place and space in collaborative systems. In: ACM (Hrsg.): CSCW '96: 6th Conference on Computer-Supported Cooperative Work. Cambridge, USA 1996, 67-76

[13]   C. Heath, P. Luff, A. Sellen: Reconsidering the Virtual Workplace: Flexible Support for Collaborative Activity. In: H. Marmolin, Y. Sundblad, K. Schmidt (Hrsg.): Proceedings of the Fourth European Conference on Computer Supported Cooperative Work (ECSCW 95). Amsterdam 1995, 83-99

[14]   S. Johnson: Interface Culture. How new technologies transforms the way we create and communicate. San Francisco 1997

[15]   D. Kiely: Are Components the Future of Software? In: Computer 2/1998, 10-11

[16]   J. Kloss u.a.: VRML 97. Bonn 1998

[17]   M. Latzer: Mediamatik: Die Konvergenz von Telekommunikation, Computer und Rundfunk. Opladen 1997

[18]   J. Lipnack, J. Stamps: Virtual Teams. New York u.a. 1997

[19]   J. Martin: Cybercorp: The new business revolution. New York u.a. 1996

[20]   R. Orfali, D. Harkey, J. Edwards: Instant CORBA. New York u.a. 1997

[21]   R. Orfali, D. Harkey: Client/Server Programming with Java and CORBA. New York u.a. 1997

[22]   A. Picot, R. Reichwald, R.T. Wigand: Die grenzenlose Unternehmung. Wiesbaden 1996

[23]   R. Reichwald, K. Möslein: Telekooperation und Dezentralisierung: Eine organisatorisch-technische Perspektive, in: K. Sandkuhl, H. Weber: ISST-Berichte Telekooperations-Systeme in dezentralen Organisationen; Tagungsband, Berlin und Dortmund 1996, 51-66

[24]   R. Reichwald u.a.: Telekooperation. Verteilte Arbeits- und Organisationsformen. Berlin, Heidelberg 1998

[25]   S. Sahay: Implementation of Information Technology: A Time-Space Perspective. In: Organization Studies 18/2 1997, 229-260

[26]   K. Schmidt: The Organization of Cooperative Work - Beyond the 'Leviathan' Conception of Organization. In: K. Schmidt, L. Bannon: Issues of Supporting Organizational Context in CSCW Systems. Lancaster 1993, 119-140

[27]   G. Schneider, J. Schweitzer: Closing the gap between synchronous and asynchronous cooperative work. In: Proc. Fifteenth International Joint Conference on Artificial Intelligence (IJCAI 97) Workshop "Business Applications of AI", NAGOYA, Aichi, Japan, August 23-29, 1997. URL: http://www-stz.dfki.uni-sb.de/~schneide/work/ijcai-97.htm, Abruf am 98-01-29

[28]   N. Streitz: Zur Zukunft computerunterstützter Gruppensitzungen. In: U. Hasenkamp (Hrsg.): Einführung von CSCW-Systemen in Organisationen. Braunschweig, Wiesbaden 1994, 225-236

[29]   D. Terpack: Konvergenz - Auf diese Gelegenheit haben wir gewartet. Firmeninformation Tektronix. URL http://www.tek.com/Measurement/TekView/TV12-1/German/convergence.html, Abruf am 98-03-04

[30]   C. Walker: Instinctive Technology ships Web-based groupware. In: PC Week Online vom 15. Oktober 1997. URL: http://www.zdnet.com/pcweek/news/1013/15meroom.html, Abruf am 98-01-12

[31]   R.T. Watson, P.G. McKeown, M. Garfield, Topologies for Electronic Cooperation. In: F. Lehner, S. Dustdar (Hrsg.): Telekooperation in Unternehmen. Wiesbaden 1997, 1-11

[32]   T. Wendel: Computerunterstützte Teamarbeit: Konzeption und Verwaltung eines Teamarbeitsystems.
         Wiesbaden 1996

Adresse des Autoren

Dipl.-Kfm. Oliver Reiss
Philipps-Universität Marburg
Institut für Wirtschaftsinformatik
35032 Marburg
E-Mail: reiss@wiwi.uni-marburg.de

Konzeption eines WWW-basierten Prozeßinformationssystems

Dr. Michael Rosemann, Boris Bachmendo
Institut für Wirtschaftsinformatik, WWU Münster

Zusammenfassung

In dem vorliegenden Beitrag wird ein Prozeßinformationssystem, ein Anwendungssystem zum Workflow-basierten Prozeßcontrolling, das die Überwachung und Auswertung der Prozeßausführung unterstützt, skizziert. Basierend auf einer kurzen Erörterung der zugrundeliegenden betriebswirtschaftlichen Motivation, werden die systemseitig zu erfüllenden Anforderungen abgeleitet. Anschließend wird exemplarisch ein mögliches Realisierungkonzept eines Web-basierten Prozeßinformationssystems anhand des Prototypen PISA 3 dargestellt. Dieser Beitrag präsentiert damit Zwischenergebnisse aus dem DFG-Projekt 'Controlling und Monitoring von verteilten Workflows zur kontinuierlichen Geschäftsprozeßoptimierung' (CONGO).

1 Verteiltes Prozeßinformationssystem – ein Workflow-basiertes Managementinformationssystem

1.1 Standortübergreifendes Prozeßmanagement als neue Herausforderung

Die Veränderung der Wettbewerbsbedingungen erhöht für viele Unternehmen die Notwendigkeit sich einer grundlegenden Reorganisation zu unterziehen, denn die Struktur des zentral geführten, funktional gegliederten Unternehmens ist den neuen Anforderungen an Flexibilität und Anpassungsfähigkeit oftmals nicht mehr gewachsen. Entsprechend gewinnen Konzepte mit alternativen Perspektiven, wie die *Prozeßorientierung*, immer mehr an Bedeutung [2]. Diese hat nicht nur in organisatorischen (z. B. Prozeßmanagement) und kostenrechnerischen (z. B. Activity-based Costing) Ansätzen ihren Niederschlag gefunden, sondern auch die Entwicklung prozeßorientierter Systeme (insb. Workflowmanagementsysteme) gefördert.

Die zunehmende Globalisierung und Internationalisierung der Märkte und die wegen des Wettbewerbsdrucks gestiegenen Anforderungen an Markt- und Kundennähe führen – begleitet durch eine wachsende Bereitschaft zur Reduktion der eigenen Leistungstiefe – überdies zur *Überwindung der Standortgrenzen*, d. h. die Geschäftsprozesse erstrecken sich zunehmend über eine größere Anzahl an Standorten. Diese Entwicklung wird durch die Entvertikalisierung der Unternehmenshierarchien und die Modularisierung der Unternehmensorganisation begünstigt [14].

Diese verstärkte geographische Verteilung der Organisationseinheiten eines Unternehmens auf der einen Seite und die zunehmende, oft nur temporäre Integration räumlich verteilter Unternehmungen (Virtuelle Unternehmen) auf der anderen führen zur Notwendigkeit der organisatorischen und funktionalen Bindung verschiedener Standorte (vgl. Abbildung 1). Als „Enabler" dienen dabei moderne Kommunikationstechnologien, insbesondere das Internet und die auf seiner Basis möglichen Dienste. Folglich kommt der Koordination standortübergreifender Geschäftsprozesse eine immer größere Bedeutung zu. Dementsprechend wächst der Bedarf an Werkzeugen, zur Konzeption, Steuerung und Kontrolle verteilter Prozesse.

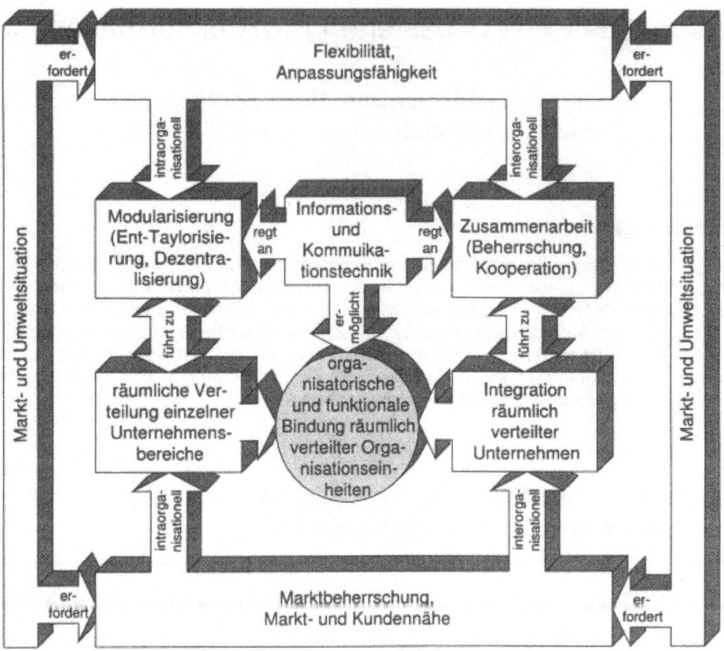

Abb. 1: Räumliche Verteilungstendenzen

1.2 Funktionalität eines Prozeßinformationssystems

Die Fokussierung auf die Geschäftsprozesse in Verbindung mit den beschriebenen Verteilungstendenzen motiviert nicht nur die Entwicklung verteilter Workflowmanagementsysteme [8, 9, 12], welche die Ausführung standortübergreifender Geschäftsprozesse unterstützen. Insbesondere bedarf es auch effizienter Managementinformationssysteme, welche auf der Basis der von Workflowmanagementsystemen aufgezeichneten und anderen operativen Informationssystemen gewonnen (Nutz-)Daten eine adressatenadäquate Informationsverdichtung und -präsentation vornehmen [20]. Eine effektive Ausnutzung der prozeßbezogenen Daten ist gerade für Virtuelle Unternehmen besonders wichtig, denn durch den häufigen Wechsel der beteiligten Partner befindet sich ein solches Unternehmen ständig am Anfang der Lernkurve [6, S. 3].

Derartige Managementinformationssysteme, die primär prozeßrelevante Daten konsolidieren und aufbereiten, werden im folgenden als *Prozeßinformationssysteme* bezeichnet (vgl. ausführlich [15]). Sie sollen durch die Informationsversorgung des Prozeßcontrolling sowohl operative als auch strategische Zielsetzungen unterstützen.

Operatives Prozeßcontrolling umfaßt die Beobachtung und Kontrolle der Prozesse bei ihrer Ausführung und wird als auch *(Prozeß-)Monitoring* bezeichnet. Aktive Monitoring-Systeme sind in der Lage Ausnahmesituationen bei der Prozeßausführung (z. B. Terminüberschreitung oder Ressourcenknappheit) zu erkennen und darauf zu reagieren (*Management-by-Exception*), indem beispielsweise verstärkt organisatorische Einheiten mit geringerer Arbeitsbelastung adressiert werden. Das passive Monitoring liefert auf Anfrage Statusinformationen über den Zustand der aktuellen Prozeßinstanzen [12, S. 67]. So bieten z. B. viele international agierende Paketdienste ihren Kunden die Möglichkeit, sich im WWW über den aktuellen Status ihrer Aufträge (sog. Package Tracking) und die zu erwartenden Lieferzeiten zu informieren.

Das *strategische* Prozeßcontrolling geht über diese Betrachtung einzelner Prozeßinstanzen hinaus, indem es auf der Basis verdichteter Daten und ggf. prozeßübergreifend über einen gewissen Zeitraum die Ablauforganisation einer Unternehmung analysiert. Das vorrangige Ziel ist die Identifizierung von Problembereichen und Optimierungspotentialen, um sie durch Prozeßreorganisation beseitigen bzw. ausnutzen zu können [7, S. 26]. Um Diskrepanzen zwischen dem tatsächlichen Verlauf und dem Prozeßentwurf festzustellen, werden die im Regelfall verdichteten Istdaten korrespondierenden Sollwerten gegenübergestellt.

Unter anderem können Prozeßinformationssysteme somit aufzeigen,

- welche Prozesse welche Durchlaufzeiten(-varianz) besitzen,

- welche Komplexitätstreiber (z. B. Prozeßtypen mit wenig Instanzen) bestehen oder

- welche Organisationseinheiten wie in die Prozesse eingebunden sind.

Die empfängeradäquat verdichteten und aufbereiteten strategischen Informationen dienen als Entscheidungsgrundlage auf unterschiedlichen Management-Ebenen und können als Input eines strategischen Entscheidungsunterstützungssystems weiterverarbeitet werden [5, S. 9ff.].

1.3 Zusätzliche Anforderungen an ein verteiltes Prozeßinformationssystem

An jedes Softwaresystem werden funktionale und nicht-funktionale Anforderungen gestellt [16]. Funktionale Anforderungen beschreiben das Verhalten des Systems und die von ihm angebotenen Dienste, während nicht-funktionale Anforderungen Bedingungen darstellen, die bei der Funktionsausführung erfüllt sein müssen. Die räumliche Verteilung führt im bezug auf Prozeßinformationssysteme zu einer Reihe zusätzlicher Anforderungen beider Arten.

Die *funktionalen* Anforderungen resultieren aus der standortübergreifenden Prozeßausführung. Bei der Betrachtung von verteilten Aktivitäten und Prozessen kommen zusätzliche Merkmale hinzu, die sich beispielsweise auf den Standort oder das Land ihrer Ausführung beziehen. Ein Prozeßinformationssystem muß demnach in der Lage sein, diese standortbezogenen Informationen zu erfassen, zu speichern und zu analysieren.

Bei der internationalen oder insbesondere bei der interkontinentalen Verteilung führen beispielsweise die bestehenden Zeitunterschiede zu einer Reihe von Problemen. Es reicht nicht aus, alle Zeitangaben beim Datenimport zu konsolidieren, auch die Unterschiede bei den Arbeitszeiten müssen berücksichtigt werden. Wenn beispielsweise eine Aktivität in Deutschland um 8 Uhr morgens abgeschlossen wurde und der darauffolgende Schritt in New York erfolgen soll, so ist eine mehrstündige Wartezeit wegen des Zeitunterschieds vorprogrammiert.

Die *nicht-funktionalen* Anforderungen sind auf die räumliche Verteilung von Mitarbeitern und Organisationseinheiten zurückzuführen. Die für das Prozeßcontrolling relevanten Informationen müssen unabhängig von der jeweiligen geographischen Position der daran interessierten Mitarbeiter zur Verfügung gestellt werden [13]. Andererseits sind die operativen Rohdaten, welche den Input eines jeden Informationsystems darstellen, auch verteilt. Für ein Prozeßinformationssystem besonders wichtig sind die Protokolle der Workflowmanagementsysteme (sog. Audit-Trail-Dateien), die sich bei verteilter Workflowsteuerung auf mehreren Workflowservern an unterschiedlichen Standorten befinden. Ein Prozeßinformationssystem muß folglich sowohl die *Verteilung der Datenquellen* als auch *der Datensenken* unterstützen.

Ein auf die Bedingungen eines verteilten Workflowmanagement zugeschnittenes Prozeßinformationssystem, das diese Anforderungen erfüllt, wird verteiltes Prozeßinformationssystem genannt. Der im folgenden skizzierte Prototyp PISA 3 ist ein System, das mit dieser Zielsetzung entwickelt wird.

2 Exemplarische Realisierung eines verteilten Prozeßinformationssystems anhand des Prototypen PISA 3

2.1 Software-Plattform

Bei der derzeit erfolgenden Implementierung des Prototyps PISA[1] 3 wurde das World Wide Web aus diversen Gründen als Basistechnologie gewählt. Zu einem kann dadurch die Erfüllung der nicht-funktionalen Anforderungen, d. h. der Umgang mit verteilten Datenquellen und -senken, gewährleistet werden. Außerdem basiert bereits die Mehrzahl der verteilten Workflowmanagementsysteme auf diesem Internet-Dienst. Als Basisdienst zur Unterstützung eines standortübergreifenden Workflowmanagements bietet das Web eine Reihe von Vorteilen:

- Durch die Übernahme der Datenaufbereitung und der Programmausführung wird das Problem der Hardwareheterogenität vollständig von dem Browser gelöst, da die WWW-Browser bereits auf praktisch allen Hard- und Softwareplattformen verfügbar sind.

- Das WWW ermöglicht einen weltweiten Zugriff auf eine rasch wachsende Datenmenge (z. B. auch auf Best Practice Cases).

- Wegen der hohen Verbreitung des WWW beherrschen die meisten potentiellen Benutzer den Umgang mit dem Browser, was zu kurzen Einarbeitungszeiten führt.

- Die lokale Installation von Client-Software entfällt, weil ihre Funktionalität komplett von dem Web-Client übernommen wird. Das Einzige, was ein Mitarbeiter für die Teilnahme am automatisierten Geschäftsprozeß oder für die Auswertung der Ablaufdaten braucht, ist ein Rechner mit einem WWW-Browser und Internet-Zugang,[2] und diese Voraussetzung ist heutzutage in den meisten Fällen erfüllt. Damit wird eine hohe Autonomie und Mobilität der Benutzer erreicht.

- Das WWW wird immer mehr zu einem wichtigen Werbe- und Vertriebsweg. Nicht nur um Kunden zu gewinnen, bietet das Web durch seine multimedialen Fähigkeiten vielseitige Möglichkeiten, auch der Kaufvorgang selbst kann z. B. mit Hilfe von Formularen automatisiert werden. Der Kauf (bzw. Vertragsabschluß) stellt den ersten Schritt vieler Verkaufsprozesse dar, und es ist naheliegend, daß nach seiner Elektrifizierung die entsprechenden Workflows automatisch angestoßen werden können.

- Workflowmanagementsysteme können oftmals in Form von aktiven (z. B. eine sich ereignisgesteuert aktualisierende Work-to-do-Liste) bzw. passiven Benutzerschnittstellen

---

[1] Der Akronym *PISA* steht für *Process Information System based on Applets*.
[2] Die Applikationen, die eventuell benötigt werden, müssen entweder lokal auf dem Rechner zur Verfügung stehen oder auch auf Web-Basis implementiert sein.

mit WWW-Browsern verwendet werden [3, 9, 10]. Letztere erlauben es beispielsweise, Workflows von außen zu instanziieren oder Monitoring-Informationen zu erhalten.

Von allen Web-basierten Implementierungstechniken zeichnet sich Java durch seine Leistungsfähigkeit, Sicherheit und Plattformunabhängigkeit (z. B. im Gegensatz zu *ActiveX* von Microsoft) aus und bietet vielseitige Möglichkeiten zur Entwicklung vollwertiger Applikationen. Die gesamte Funktionalität von PISA 3 und die graphische Benutzeroberfläche soll durch Java-Applets implementiert werden, wodurch eine hohe Verfügbarkeit gewährleistet ist. Durch verschiedene Authentifizierungsmechanismen kann der Zugriff auf einen vordefinierten Personenkreis eingeschränkt werden. Wenn ein Benutzer regelmäßig von einem bestimmten Rechner aus mit PISA arbeitet, sollen die Programm-Dateien (d. h. Class-Dateien, die den vorkompilierten Code enthalten) in einem speziellen Verzeichnis des Browsers installiert werden, so daß ihre zeitaufwendige Übertragung entfällt.

Als Datenbasis dient für PISA das *Microsoft SQL-Server* Datenbankmanagementsystem, das auf dem WWW-Server von PISA installiert ist. Der Zugriff der auf dem Client ablaufenden Applets auf die Datenbank erfolgt über JDBC.[3] JDBC ist eine Java API, die das Erstellen einer Datenbankverbindung über das Internet, das Senden von SQL-Statements an diese Datenbank und das Empfangen von Ergebnissen erlaubt.

Bei der Realisierung von Datenbankanwendungen auf Web-Basis sind unterschiedliche Architekturvarianten denkbar [1]. PISA basiert auf der Drei-Schicht-Architektur, die aus Client-Anwendungen, Datenbankservern und einer Mittelschicht besteht (vgl. Abbildung 2). Die Mittelschicht bildet der dbANYWHERE Server von Symantec Corporation [18], der als Datenbankagent betrachtet werden kann. Der Client stellt lediglich eine Verbindung zum dbANYWHERE Server her, der seine Anfragen an die Datenbank und die Ergebnisse an ihn zurückleitet. Da auf dem Client keine datenbankspezifische Software installiert wird, gewährleistet diese Architektur die Datenbanktransparenz, d. h. die Datenbank kann ausgetauscht oder auf einen anderen Rechner verlegt werden, ohne Anpassungen auf der Client-Seite zu erfordern.

Die drei Architekturschichten können auf unterschiedlichen Rechnern installiert sein oder beliebig zusammengefaßt werden. Bei PISA soll sowohl die Datenbank als auch der dbANYWHERE-Server auf dem als WWW-Server agierenden Rechner installiert werden, der im folgenden als PISA-Server bezeichnet wird.

---

[3] JDBC ist ein eingetragenes Warenzeichen von Sun Microsystems und kein Akronym, trotzdem wird es oft als *Java Database Connectivity* interpretiert [17].

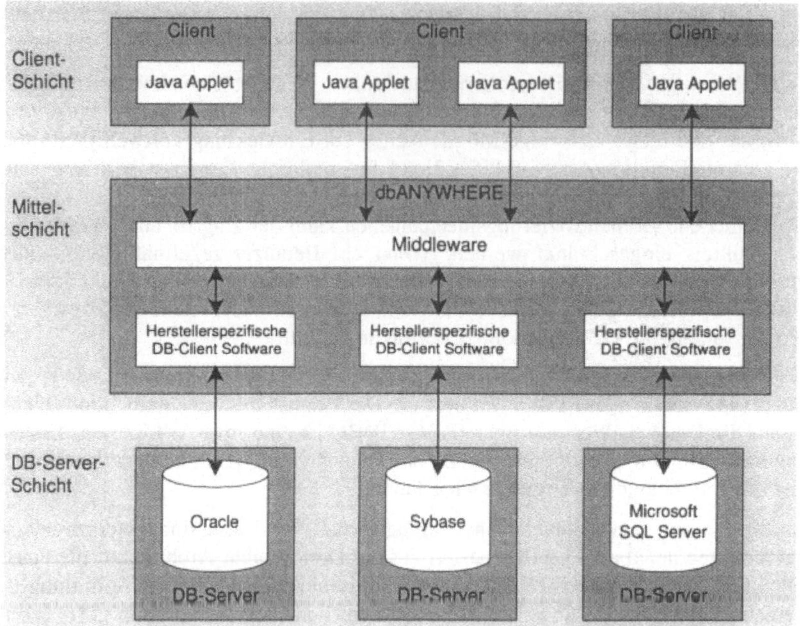

Abb. 2: Drei-Schichten Architektur mit dbANYWHERE

2.2 Nicht-funktionale Eigenschaften

Durch die Nutzung der WWW-Technologie entstehen vielfältige Möglichkeiten für den Einsatz des Prozeßinformationssystems. Bei der Installation auf einem an ein Intranet angeschlossenen Server, kann PISA von allen Organisationseinheiten unabhängig von ihrer geographischen Position benutzt werden. Damit trägt das System beispielsweise dezidiert den Anforderungen Virtueller Unternehmen Rechnung.

Dank der Drei-Schichten Architektur kann das System mit einer beliebigen Datenbank verbunden werden. Dafür muß bei dbANYWHERE der entsprechende Treiber vorhanden und die Internet-Adresse des Datenbank-Servers bekannt sein. Im folgenden werden die PISA nutzenden Unternehmen als *Kunden* und ihre Mitarbeiter als *Benutzer* bezeichnet. Beim kommerziellen Einsatz von PISA wäre es entsprechend denkbar, daß es nicht verkauft, sondern im Internet zur Verfügung gestellt wird, und für seine Nutzung Gebühren (z. B. nach Zeit) erhoben werden.

Die Daten verschiedener Kunden müssen dabei getrennt gespeichert werden. Die Trennung der Datenbestände ist auf zwei Ebenen realisiert. Ein Kunde kann seine Daten entweder mit Hilfe des internen Datenbankmanagementsystems auf dem PISA-Server speichern oder sein eigenes (externes) DBMS benutzen (vgl. Abbildung 3). Letzteres kann, z. B. aus Sicherheitsgründen, bevorzugt werden. Insbesondere muß dabei der Datenimport nicht mehr über das Internet erfolgen und die Importmodule können vom Kunden selbst implementiert werden. Auch für den Import in die interne Datenbank ist die Benutzung kundeneigener Systeme möglich. Im Rahmen von PISA ist die Implementierung der Importmodule für einige verbreitete Datenquellen (z. B. das Workflowmanagementsystem IBM-FlowMark und das Modellierungstool ARIS-Toolset) geplant.

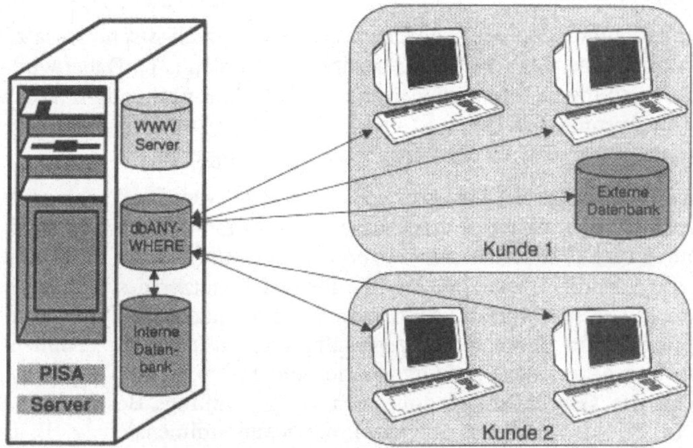

Abb. 3: Einsatzmöglichkeiten von PISA

Jedes Datenbankmanagementsystem enthält mehrere physisch oder logisch getrennte Datenbasen (bzw. Schemata). Auch auf dem SQL-Server von PISA kann ein Kunde mehrere Datenbasen erstellen und benutzen. Sie können dazu benutzt werden, die Daten nach einem bestimmten Kriterium (z. B. Zeitraum oder Region) aufzuteilen. Nachdem PISA von der Web-Seite gestartet wurde, erscheint das Database-Connection-Fenster, in dem Einstellungen für die Verbindung mit der internen oder externen Datenbank vorgenommen werden.

Nach der Erstellung der Datenbank-Verbindung erfolgt die Benutzeranmeldung. Jeder Benutzer von PISA hat ein eindeutiges Login und Paßwort. Die Benutzerrechte werden auf zwei Ebenen vergeben: auf Systemebene wird festgelegt, welche Systemfunktionen der User benutzen kann, auf Datenebene wird vorgegeben, welche Rechte er auf welche Daten besitzt.

Auf der *Systemebene* wird jeder Benutzer einer der drei Gruppen zugeordnet:

- der *Evaluator* darf die Daten nur auswerten,
- der *Manager* darf sie außerdem verwalten (d. h. importieren und löschen),
- der *Administrator* hat alle datenbezogenen Rechte und kann zusätzlich die Benutzer verwalten (d. h. neue anlegen, vorhandene löschen und ihre Rechte ändern).

Auf der *Datenebene* erfolgt die Vergabe von Rechten über Domains. Eine *Domain* umfaßt mehrere Workflow-Server, wobei jeder Server auch mehreren Domains zugeordnet werden kann. Die Domains werden von Kunden nach beliebigen Kriterien gebildet (z. B. Standort). Ein Benutzer darf aus einer Domain stammende Daten entweder nur auswerten, oder auswerten und verwalten, d. h. er ist entweder Evaluator oder Manager. Wenn für eine Benutzer-Domain-Konstellation kein Status spezifiziert ist, hat der Benutzer keinen Zugriff auf die entsprechenden Daten. Bei Berechtigungsdiskrepanz auf den beiden Ebenen gilt immer die strengere Vorgabe. So kann z. B. einem Systemadministrator der Zugriff auf bestimmte Daten verwehrt werden.

2.3 Funktionale Eigenschaften

Zur Analyse der Prozeßinformationen sollen bei PISA sowohl deduktive als auch induktive Verfahren realisiert werden.

Deduktive oder *hypothesenüberprüfende* Datenanalyse ist der klassische Ansatz, bei dem Benutzeranfragen vom System bearbeitet und beantwortet werden [21]. Dabei wird die Datenbasis gezielt nach bestimmten Informationen durchsucht. Eine notwendige Voraussetzung für eine effektive Auswertung von Daten ist ihre geeignete Strukturierung, die es dem Benutzer erlaubt, auf optimale Weise durch den Datenbestand zu navigieren.

E. F. CODD et al. bezeichnen die Fähigkeit „to consolidate, view, and analyze data according to multiple dimensions, in ways that make sense to one or more specific enterprise analysts at any given point of time" [4] als *multidimensionale Datenanalyse*. Bei PISA werden die Auswertungsobjekte (d. h. die Umweltobjekte, die für den Benutzer des Systems von Interesse sein können) zu den sechs folgenden Dimensionen zusammengefaßt, wobei die semantische Zusammengehörigkeit aus Benutzersicht zugrundegelegt wird: die Prozeßdimension (sie umfaßt Workflows und Aktivitäten), die Organisationsdimension (z. B. Mitarbeiter und Organisationseinheiten), die Geschäftsobjektdimension (z. B. Auftrag, Bestellung), die Ressourcendimension (z. B. Hardware und Software), die Standortdimension (z. B. Land, Werksstandort) und die Zeitdimension (Zeitpunkt, Zeitraum).

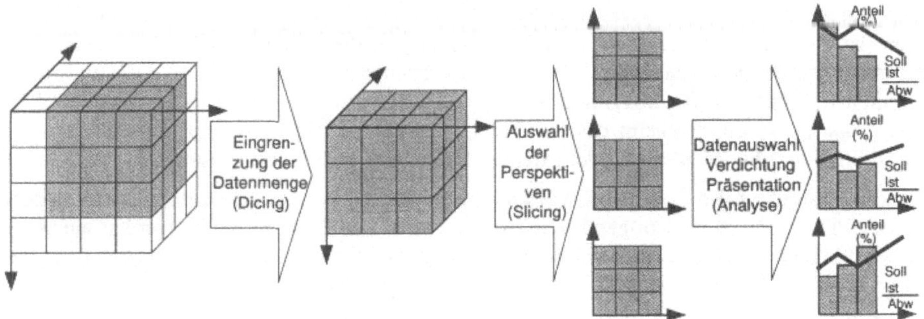

Abb. 4: Auswertungsschritte

Die PISA zugrundeliegende Vorgehensweise bei der Datenauswertung ist in der Abbildung 4 schematisch dargestellt. Im ersten Schritt wird die Datenmenge entlang allen Dimensionen auf relevante Daten eingeschränkt (dies wird als *Dicing* bezeichnet). Im nächsten Schritt werden die Perspektiven gebildet (*Slicing*), d. h. der durch Dicing entstandene Würfel wird seitenweise betrachtet. Eine Perspektive entspricht einer Seite des Würfels und umfaßt zwei Dimensionen. Wenn beispielsweise die Lernkurve eines Mitarbeiters in bezug auf einen bestimmtes Workflow dargestellt werden soll (die Perspektive besteht aus Organisations- und Prozeßdimensionen), werden alle Instanzen des Modells über alle Zeiträume, Standorte oder Ausführungsrechner hinweg betrachtet (innerhalb des durch Dicing entstandenen „Unterwürfels"). In jeder Perspektive werden im dritten Schritt die ausgewählten Daten verdichtet und auf eine geeignete Weise präsentiert.

Im Gegensatz zum deduktiven Ansatz, wird bei der *induktiven* Datenanalyse die *Hypothesenerzeugung* automatisiert, d. h. das System analysiert die Datenmenge und weist den Benutzer auf die darin enthaltenen Zusammenhänge hin. Das Vorgehen soll anhand der *Clusteranalyse* erläutert werden. Bei diesem Verfahren werden die Datenobjekte anhand verschiedener Merkmale in Gruppen (Cluster) zusammengefaßt, um so auf die Zusammenhänge zwischen bestimmten Merkmalsausprägungen schließen zu können. Beispielsweise können Rechnungen sowohl nach Lieferanten als auch nach ihrer Korrektheit geclustert werden. Wenn beispielsweise der Anteil eines bestimmten Lieferanten an der Gruppe der zu reklamierenden Rechnungen höher ist, als an der Grundgesamtheit, ist dies ein Zeichen für bestehenden Handlungsbedarf (z. B. Anpassung der lieferantenspezifischen Konditionen).

3 Résumé und Ausblick

Moderne Informations- und Kommunikationstechnologien stellen nicht nur die Werkzeuge für die Kooperation zwischen mehreren Standorten zur Verfügung, sondern fördern durch die Entwicklung immer benutzerfreundlicherer und leistungsfähigerer Kommunikationswege die bereits betriebswirtschaftlich motivierten ('Konzentration auf die Kernkompetenzen') Verteilungstendenzen. Diese enge Verflechtung zwischen der Entwicklung der betriebswirtschaftlichen Theorie und dem informationstechnischem Fortschritt wird am Beispiel des standortübergreifenden, Workflow-basierten Geschäftsprozeßmanagement deutlich [20]. Die softwaretechnische Unterstützung darf sich jedoch nicht alleine auf die Unterstützung der (Web-basierten) Kommunikation beschränken, sondern muß auch zur Koordination der Geschäftsprozesse wesentliche Beiträge leisten.

Dieser Zusammenhang hat auch die Entwicklung des Prozeßinformationssystems PISA maßgeblich geprägt. Die ersten beiden am Lehrstuhl für Wirtschaftsinformatik und Informationsmanagement der Universität Münster entwickelte Prototypen wurden mit Hilfe von *MS-Access* implementiert und konnten insbesondere die im Kapitel 1.3 beschriebenen nichtfunktionalen Anforderungen nur unzureichend erfüllen. Erst nachdem durch die Weiterentwicklung von Java und JDBC die Implementierung vollwertiger GUI-gesteuerter verteilter Anwendungen mit Internet-weitem Datenbankzugriff möglich wurde, kann der WWW-basierte Prototyp PISA 3 realisiert werden.

Zur Zeit wird an der Weiterentwicklung des Systems mit Java 1.1 gearbeitet. Es wird ferner die Möglichkeit erwogen, *OLAP-Tools* einzusetzen, welche die multidimensionalen Datenstrukturen schon auf der Datenbankebene realisieren bzw. simulieren, was zur erheblichen Leistungssteigerung bei den Datenbankzugriffen führt (einen Überblick über die verfügbaren Systeme bietet z. B. [19]). Ein weiterer Aspekt, der im Rahmen der Weiterentwicklung behandelt werden soll, ist der Datenschutz. Die Sicherheit der Daten bei ihrem Import in die Datenbank oder bei der Übertragung von der Datenbank zum Benutzer kann beispielsweise durch Secure Sockets Layer (SSL) gewährleistet werden. Darüber hinaus soll die Interaktion mit einem ebenfalls im WWW verfügbaren Prozeßmodellierungstool (z. B. ARIS-Toolset in Verbindung mit dem ARIS Internet Navigator) realisiert werden.

Adressen der Autoren

Dr. Michael Rosemann
Westfälische Wilhelms-Universität Münster
Institut für Wirtschaftsinformatik
Steinfurter Str. 107
48149 Muenster
E-Mail: ismiro@wi.uni-muenster.de

Boris Bachmendo
Westfälische Wilhelms-Universität Münster
Institut für Wirtschaftsinformatik
Steinfurter Str. 107
48149 Muenster
E-Mail: isboba@wi.uni-muenster.de

Literaturverzeichnis

[1]   W. Benn, I. Gringer: Zugriff auf Datenbanken über das World Wide Web. In: Informatik-Spektrum 21 (1998), 1, 1-8.

[2]   S. Boyd: Process-Driven Workflow. In: L. Fischer (Hg.): The Workflow Paradigm. The Impact of Information Technology on Business Process Reengineering. 2nd Ed., Future Strategies, Lighthouse Point, Florida 1995, 15-24.

[3]   Chr. Bußler et al.: Das WWW als Benutzerschnittstelle und Basisdienst zur Applikationsintegration für Workflow-Management-Systeme. In: EMISA-Forum 6 (1997), 1, 85-90.

[4]   E. F.Codd, S. B. Codd, C. T. Sally: Providing OLAP (On-Line Analytical Processing) to User-Analysis: An IT Mandate. 1993.
        http://www.arborsoft.com/essbase/wht_ppr/coddTOC.html 09.03.1998.

[5]   W. Deiters, R. Striemer: Prozeßmanagementsysteme – Basistechnologie für ein entscheidungsorientiertes Informationsmanagement. Fraunhofer-Institut für Software- und Systemtechnik, Bericht 23. Dortmund 1995.

[6]   W. Faisst: Wissensmanagement in Virtuellen Unternehmen. In: D. Ehrenberg, J. Griese, P. Mertens (Hg.): Arbeitspapier der Reihe „Informations- und Kommunikationssysteme als Gestaltungselement Virtueller Unternehmen", Nr. 8/1996 Nürnberg 1996.

[7]   J. Galler, A.-W. Scheer: Workflow-Projekte: Vom Geschäftsprozeßmodell zu unternehmensspezifischen Workflow-Anwendungen. In: Information Management 10 (1995), 1, 20-27.

[8]   D. Georgakopoulos, M. Hornick, A. Sheth: An Overview of Workflow Management: From Process Modeling to Workflow Automation Infrastructure. In: Distributed and Parallel Databases 3 (1995), 2, 119-153.

[9]   H. Groiss, J. Eder: Kooperation von Workflowsystemen im World-Wide-Web. In: EMISA-Forum 6 (1997), 1, 90-95.

[10]   A. Grasso, J-L. Meunier, D. Pagani, R. Pareschi. Distributed Coordination and Workflow on the World Wide Web. In: Computer Supported Cooperative Work 6 (1997), 2-3, 175-200.

[11]   F. Leymann, W. Altenhuber: Managing business processes as an information resource. In: IBM Systems Journal 33 (1994), 2, 326-348.

[12]   P. Koksal, S. N. Arpinar, A. Dogac: Workflow History Management. In: Sigmod Record, 27 (1998), 1, 67-75.

[13]   M. Müller-Wunsch: Verteilte Führungsunterstützungssysteme mit hybriden Problemlösungsfähigkeiten. In: M. Klotz, H. Wenzel (Hg.): Führungsinformationssysteme im Unternehmen: Erfolgsfaktoren, Vorgehensweisen und Perspektiven. Berlin, 1994, 193-210.

[14]   A. Picot, R. Reichwald, R. T. Wigand: Die grenzenlose Unternehmung: Information, Organisation und Management. 2. Auflage. Gabler, Wiesbaden, 1996.

[15]   M. Rosemann: Arbeitsablauf-Monitoring und -Controlling. In: St. Jablonski, M. Böhm, W. Schulze (Hg.): Workflow-Management. Entwicklung von Anwendungen und Systemen. dpunkt, Heidelberg, 1997, 201-210.

[16]   I. Sommerville: Software Engineering. 5th Ed. Bonn et al., 1996.

[17]   Sun Microsystems: JDBC Guide: Getting Started.
        http://java.sun.com/products/jdk/1.1/docs/guide/jdbc/getstart/introTOC.doc.html. 09.03.1998.

[18] Symantec Corporation: Symantec dbANYWHERE for Windows NT 3.51 & 4.0. Preview Release 3. 1996.

[19] L. Thé: OLAP Answers Tough Business Questions. In: Datamation o. Jg. (1995), 1, 65-72.

[20] G. Weikum: Workflow Monitoring:Queries On Logs or Temporal Databases? Proceeings of High Performance Transaction Systems (HPTS) Workshop 1995. http://www3.hursley.ibm.com/hpts95/proc95.htm 09.03.1998.

[21] J. Zytkov, J. Baker: Interactive Mining of Regularities in Databases. In: G. Piatetsky-Shapiro, W. J. Frawley (Hg.): Knowledge Discovery in Databases. AAAI Press, Menlo Park et al., 1991, 31-54.

Internet-Telefonie: Chancen und Risiken

Wulf Bauerfeld

Deutsche Telekom Berkom GmbH, Berlin

"Internet-Telefonie" - klassische Telefon-Dienste wie Telefonieren oder Fax werden nicht über herkömmliche Telefonnetze, sondern über IP-Technologie erbracht - ist in vieler Munde. Internet-Telefonie ist - ungeachtet des Namens - auch in geschlossenen Unternehmensnetzen (Intranets) möglich, wenn diese auf dem Internet-Protokoll IP basieren.

Damit (analoge) Sprachsignale über ein Datennetz verschickt werden können, werden sie mit Hilfe eines besonderen Programmpaketes (Telephony-Software) digitalisiert und in entsprechende IP-Pakete aufgeteilt (Paketierung). Die Pakete erhalten außerdem die erforderliche Zieladresse im Netzwerk, bevor sie auf die Reise geschickt werden. An der Zieladresse angekommen, werden die digitalen Daten wieder in Sprache umgewandelt.

1. Internet-Telefonie - wie funktioniert das

Dieses Prinzip kann auf unterschiedliche Art und Weise umgesetzt werden - abhängig davon, ob die Anwender PCs mit Netzanschluß oder herkömmliche Telefone benutzen.

1.1 Von PC zu PC mit festem Internet-Anschluß

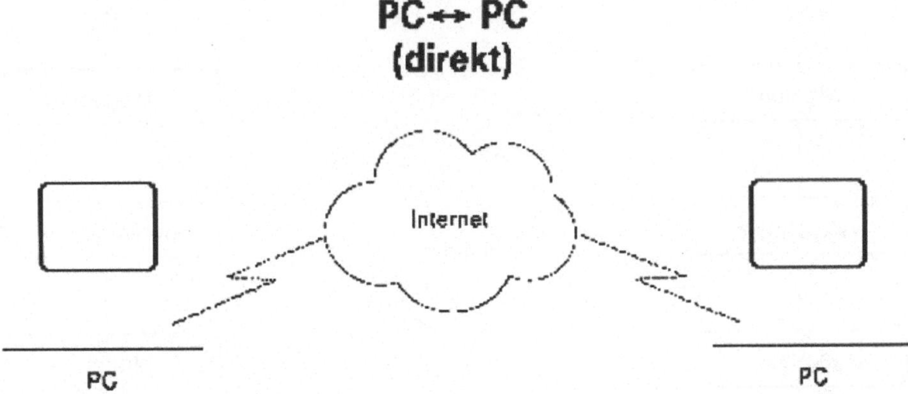

Beide Gesprächsteilnehmer haben einen PC mit festem Anschluß an das Internet ("IP-dial tone"). Die PCs sind ausgerüstet mit Soundkarte, Lautsprechern und Mikrofon und Telefony-Software, deren Kernstück das **"Speech Compression System"** ist (s.Kasten).

Z.Zt. gibt es fast vierzig verschiedene Software-Produkte, die nur sehr bedingt miteinander kompatibel sind; eine Anpassung aller Produkte an den Standard H.323 wird aber in spätestens zwei Jahren abgeschlossen sein.

Ein "IP-dial tone" Anschluß, über den jederzeit die Kommunikation über Internet aufgenommen werden kann, ist fast ausschließlich auf die Geschäftswelt beschränkt und wird meist über LAN (sozusagen als Internet-Untervermittlung) realisiert. Damit ist auch dem Teilnehmer eine feste IP-Adresse zugewiesen, über die Gesprächspartner immer erreicht werden können (vorausgesetzt, der PC ist eingeschaltet).

Im privaten Bereich wird erst die Anbindung über xDSL (über die Telefonleitung) oder über ein Kabelmodem (über das Fernseh-Kabelnetz) die Installation einer Internet-Steckdose ermöglichen und damit einen wirklichen Durchbruch auch für andere IP-basierte Anwendungen erzielt werden. Bis dahin wird man sich über das Telefonnetz zu einem "Internet -Provider" einwählen und sich mit einer dynamisch zugewiesenen IP-Adresse zufrieden geben müssen.

1.2   Von PC zu PC über lokale Internet-Provider

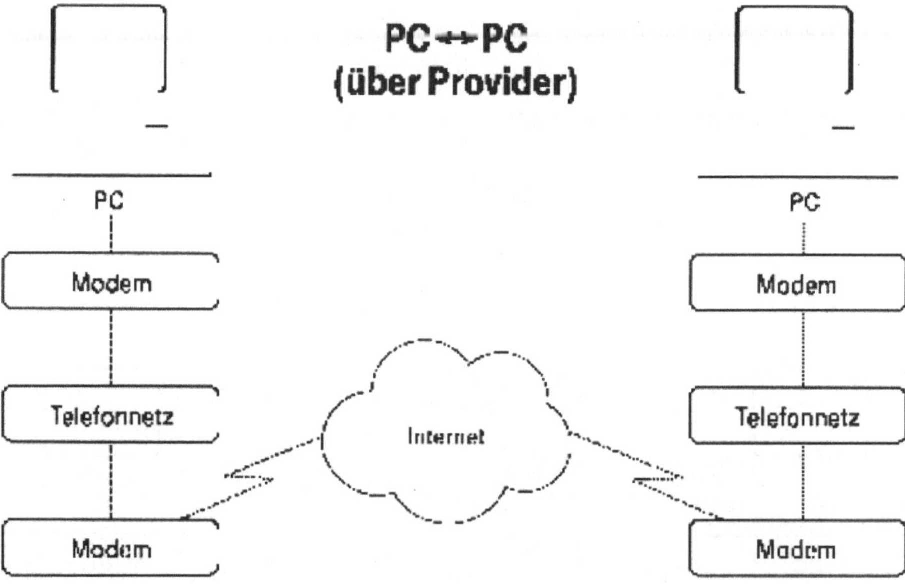

Hier haben die Teilnehmer keinen festen Internet-Zugang , vor der Gesprächsaufnahme über Internet muß erst die "dynamische Adresse" des potentiellen Gesprächspartners erfragt (z.B. über das konventionelle Telefon) werden.

Dennoch steckt in dieser Konfiguration der Erfolgsschlüssel zur Internet-Telefonie, deren grundsätzliche Spezifikationen ("telephony over packet networks") schon vor mehr als fünfzehn Jahren festgelegt wurden. Erst aber, als Übertragungsgeschwindigkeiten der Modems und die von Kompressionsalgorithmen benötigte Bandbreite einander "trafen", gelang einer israelischen Firma mit einer entsprechenden Telephony-Software für den 3.86-Prozessor unter "Windows" der entscheidende Durchbruch.

1.3 Von Telefon zu Telefon

Für diese Variante der Internet-Telefonie müssen beide Gesprächspartner lediglich ein tonwahlfähiges Telefon besitzen. Allerdings muß auf beiden Seiten für die Verbindung zwischen Internet und Telefonnetz gesorgt werden. Diese Konstellation benötigt daher als Bindeglied einen speziellen Dienstleister, der über eine besondere technische Einrichtung (Gateway genannt und mit "Speech Compression System" ausgestattet) verfügen muß. Das Gateway ermöglicht es, Sprache aus dem Telefonnetz zu digitalisieren und im Internet weiterzuleiten bzw. Daten aus dem Internet in Sprache zu verwandeln und in das angeschlossene Telefonnetz einzuspeisen. Mit dieser Lösung können zwei nationale Telefonnetze über das Internet miteinander verbunden werden: Jeder Teilnehmer in einem der beiden nationalen Telefonnetze kann alle Teilnehmer des anderen Netzes erreichen und umgekehrt.

1.4 Vom PC zum Telefon oder umgekehrt

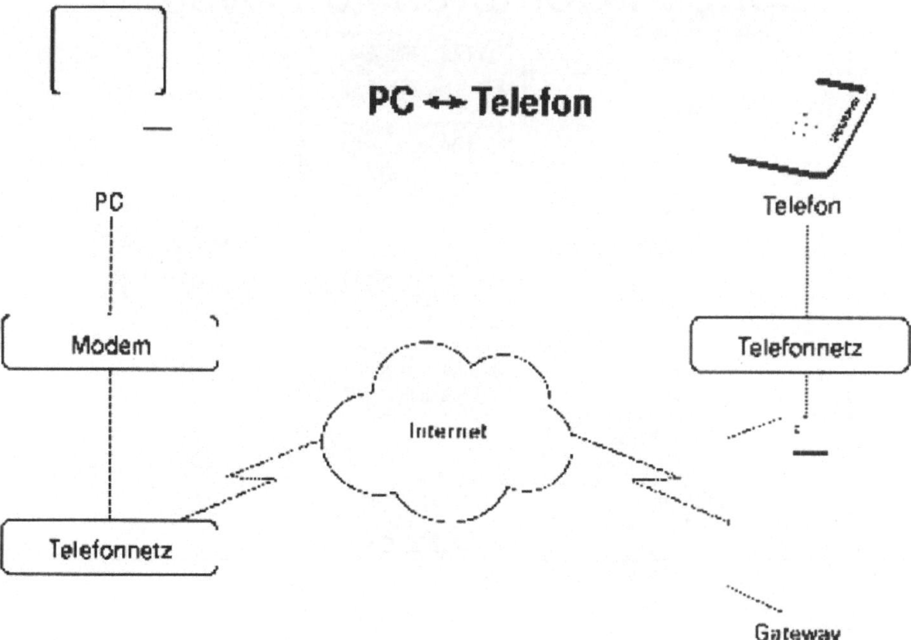

Bei dieser Variante hat ein Teilnehmer einen PC mit Netzverbindung, Soundkarte, Lautsprecher und Mikrophon. Sein Gesprächspartner hat ein tonwahlfähiges Telefon.

Bei dieser Integration von Telefonnetz und Internet offenbaren sich die grundsätzlichen Probleme der Telefonie über unterschiedliche Technologien:

- übliche Dienstmerkmale wie Rufnummernanzeige, Wahlwiederholung im "Besetztfall", Rufumleitung oder Konferenzschaltung können (noch) nicht übertragen werden

- es gibt keinen einheitlichen Rufnummern-Plan

- PCs sind nicht immer "online" und beim Internetzugang über Einwahl werden IP-Adressen meist dynamisch vergeben

Erst wenn Internet-Telefonie-Gateways einerseits das sehr komplexe Telefonie-Signalisierungssystem SS 7 nicht nur "verstehen" sondern auch übersetzen können und andererseits PCs über einen "IP dial tone" ständig erreichbar sein werden, wird sich "Internet-Telefonie" zwar nicht als scheinbar kostengünstigere Alternative, aber als Integrationsinstrument zwischen Sprach- und Datennetzen im öffentlichen Bereich wirklich durchsetzen.

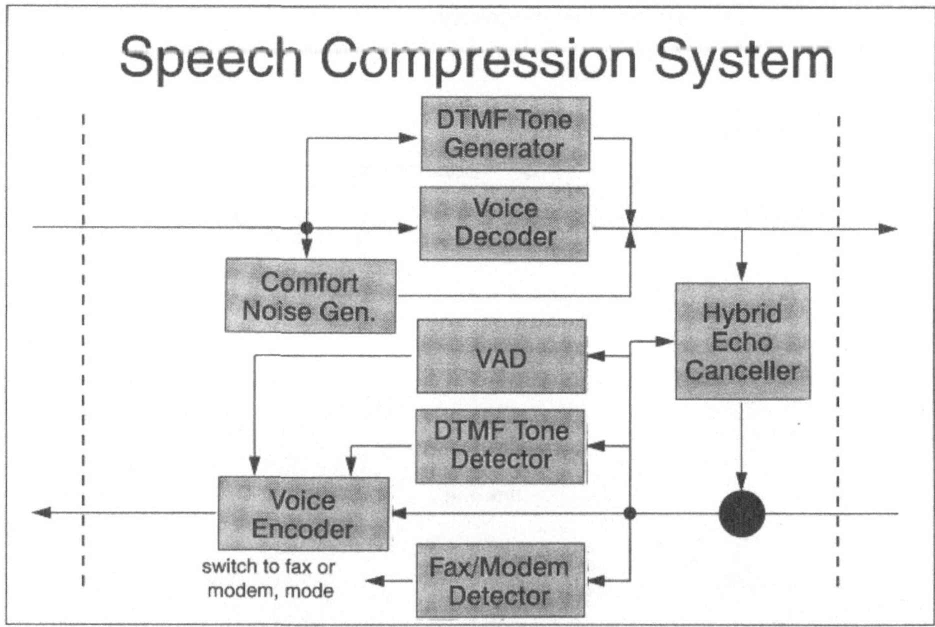

Voice Encoder/Decoder:	**G. 723.1 (6.4 kb/s oder 5.3 kb/s) ist einer von vielen Standards, die z. Zt. implementiert werden**
DTMF Tone Detector/Generator:	**Töne werden vom sprach-optimierten System nur schlecht übertragen, sie sollten in eigenständige IP-Pakete umgewandelt werden**
Fax/Modem Detector	**FAX- oder Modem-"Töne" dürfen nicht komprimiert und müssen gesondert behandelt werden**
Voice Activity Detector	**Wird "Schweigen" nicht übertragen, läßt sich die Anzahl der IP-Pakete um den Faktor 2 verringern**
Comfort Noise Generation	**"Schweigen" des Gesprächspartners sollte durch ein leises "Rauschen" ersetzt werden**
Echo Canceller	**"Echo-Kompensation" ist zwingend, um einen gewissen Gesprächskomfort zu erreichen**

2. Von Deutschland in die USA zum Ortstarif?

Die oft geäußerte Aussage, Internet-Telefonie ermögliche weltweite Ferngespräche zum Ortstarif, kann in dieser Form nicht aufrecht erhalten werden, wie folgende Tabelle zeigt.

2.1 Kostenfaktoren der Internet-Telefonie

	Zugangsnetz A / Telefon/ISDN	Internet	Zugangsnetz B / Telefon/ISDN
Von PC zu PC mit direktem Internet-Anschluß	-	x	-
Von PC mit Wählan- schluß zum Internet zu PC (mit Wählan- schluß zum Internet) oder zu Telefon	x	x	x
Von Telefon zu Telefon	x	x	x

In fast allen Szenarien fallen zusätzlich zu den Kosten für die Verbindung im lokalen Netz weitere Kosten für die Übertragung im Internet sowie für die Nutzung des Fernsprechnetzes auf der Gegenseite an. Nur im ersten Szenario, das derzeit nur für wenige Nutzer in Frage kommt, bleiben die Kosten auf die Datenübertragung im Internet begrenzt.

Wegen der unterschiedlichen Tarifmodelle und Abrechnungsarten für Internet-Bandbreiten können die hierfür anfallenden Kosten nicht ohne weiteres mit den Tarifen für lokale Telefongespräche verglichen werden.

2.2 Ein Pilotversuch von Telefon zu Telefon

Als internationalen Pilotversuch startete die Deutschen Telekom im Spätsommer letzten Jahres "T-NetCall". Ziel des Versuches war es, die Möglichkeiten und die Akzeptanz des Telefonierens über das Internet zu erkunden. Der Pilotversuch hatte eine Laufzeit von drei Monaten.

Am Pilotversuch T-NetCall nahmen 1.000 Anwender aus Deutschland, Japan, Kanada und den USA teil. Die Möglichkeit, das Internet für Ferngespräche zu nutzen, war im Rahmen des Pilotversuchs auf Verbindungen zwischen diesen vier Ländern und dort auf bestimmte Regionen beschränkt. Die Akzeptanz des neuen Dienstes wurde durch eine begleitende Umfrage unter den Pilotanwendern ermittelt.

Für das Telefonieren über das Internet benutzten hier die Teilnehmer ihr gewohntes Telefon. Als einzige Voraussetzung muß dieses Telefon auf Tonwahl eingestellt werden. Nach Anwahl einer zentralen Rufnummer im Inland wurden die Teilnehmer per Ansage aufgefordert, die Zielnummer im Ausland und ihre persönliche Geheimzahl (PIN) auf der Telefontastatur einzugeben. Die komplexe Technik blieb den Nutzern zwar verborgen, mögliche Ärgerniß für die Dauer des Gesprächaufbaus wurde mit der Ansage "Ihr T-NetCall wird aufgebaut" gemildert. .

Die im Projekt T-NetCall eingesetzte Technik ermöglicht das direkte Gespräch von Telefon zu Telefon. Für das Pilotprojekt wurden in Deutschland, Japan, Kanada und den USA Gateways eingerichtet, die über normale Telefone erreichbar waren. Sie übernahmen die Umwandlung von Sprache in Daten und deren Weiterleitung über das Internet in das Telefonnetz des Gesprächspartners.

Dienste wie T-NetCall sind von der Produktgestaltung her vergleichbar mit einem "Call-by-Call"-Angebot: über eine bestimmte Rufnummer erreicht man einen Dienstanbieter der dem über eine Geheimzahl identifizierten Kunden nationale oder internationale Fernverbindungen anbietet. Die "junge" IP-Technologie ist nicht erwiesenermaßen billiger, weder in den Investitions- noch in den Betriebskosten. Auch in der herkömmlichen Telefonie lassen sich Sprachsignale bis unter 6kbit/s komprimieren und so auf einer ISDN-Verbindung ca. 10 Gespräche bündeln.

Aber: In den meisten Ländern der Welt unterliegt Internet-Telefonie (noch) keiner Regulierung und es bedarf keiner besonderen Lizenz sich als Internet Service Provider (ISP) oder als Internet Telephone Service Provider (ITSP) aufzustellen.

Daher sind auch eine Reihe von "Next Generation Telecommunication Operators - NextGen-Telcos" entstanden, die insbesondere einen Nischenmarkt für ethnische Minderheiten (z.B. Gespräche zwischen New York und Israel oder zwischen deutschen Großstädten und der Türkei) entdeckt haben. Im Vergleich zu den großen "Carriers" sind Anzahl der Kunden und getätigten Umsätze aber eher bescheiden. Außerdem erlaubt die heute auf basierende Gateway-Technik auch noch keinen Ausbau bis zur "Carrier-size".

In naher Zukunft wird sich die IP-Telefonie von Telefon zu Telefon als öffentliches Angebot daher ausschließlich über den Preis in Relation zum konventionellen Telefonnetz und die Einfachheit der Bedienung durchsetzen müssen.

Anders wird es in firmeneigenen Sprach-/Datennetzen aussehen, wenn das Datennetz auf IP umgestellt wurde oder wird. Hier kann es zu erheblichen Kosteneinsparungen führen, wenn insbesondere bei Fernverbindungen IP-Telefonie als kleiner Teil eines sonst erheblich größeren Datenverkehrs im Intranet mitlaufen kann.

2.3   Von PC zu PC mit direktem Internetanschluß

Dort wo PCs fest am Internet oder einem Intranet angeschlossen sind (in Firmen und in Universitäten), fehlt in der Regel die zusätzliche Ausstattung mit einem Soundboard bzw. mit Mikrofon oder Lautsprecher. Auch wenn auf jedem Arbeitsplatz einer Firma ein PC stehen würde (in Deutschland haben ca. 35 % aller Arbeitsplätze zumindest Zugang zu einem Rechner), bleibt zur Zeit ein Szenario doch recht unwahrscheinlich, daß in solchen Firmen alle Telefonapparate abgeschafft und PCs so aufgerüstet werden, daß sämtliche internen Telefonate über PCs und externe Gespräche über einen Gateway geführt werden.

Eine interessante Alternative einer solchen "direkten" IP-Telefonie stellen aber die mittlerweile auf dem Markt verfügbaren IP-Telefonapparate dar. Diese sehen aus wie ein "normaler" Telefonapparat, werden bedient wie ein "normaler" Telefonapparat, haben im Inneren aber ein Speech Compression System, erzeugen oder empfangen IP-Pakete und werden an eine IP-basiertes Netz (z.B. ein Ethernet-LAN) angeschlossen. Koordiniert werden alle IP-Telefonate über einen zentralen Server im Inter(sub-)net.

2.4   Von PC zu PC über lokale Internet-Provider

Anders sieht es aus bei den privaten oder beruflichen "Web-Surfern", also den Kunden eines Online-Dienstes, die sich in der Regel bei diesem Internet Service Provider einwählen.

Hier sind die PCs als potentielle "Spielmaschinen" bereits mit Soundboard und Lautsprechern, vielleicht sogar mit Mikrofon oder einem "Headset" ausgestattet. Die grundsätzlichen Schwierigkeiten der Erreichbarkeit eines Partners gelten zwar weiterhin für alle Einwahlkunden, nicht aber für die Web-Server von Firmen, die mit einem festen Anschluß ihre Präsenz im Internet zeigen. Oft betreiben (oder lassen betreiben) solche Firmen auch ein Call-Center, welches eine "Hot-Line" bedient oder einen 24-Stunden offenen Bestell-Dienst oder andere Kundendienste anbietet.

Folgerichtig werden im Internet-Vorreiterland USA viele Web-Seiten durch einen Bedienknopf ergänzt, der ein Aufforderung zur direkten Aufnahme eine IP-Telefonats auffordert. Wählt man diese Symbol an wird zwischen dem "Internet-Surfer" und dem Web-Server eine IP-Telefonie-Verbindung aufgebaut, die über einen Gateway an die konventionellen Telefone des Call-Centers weitergeleitet wird.

Call-Center werden in den USA üblicherweise über eine (für den Kunden) entgeltfreie 800-Nummer erreicht, für die Gesprächskosten muß dann das Call-Center aufkommen. Diese sehen daher in ihrer Erreichbarkeit über Internet ein großes Einsparungspotential.

3.   Potentiale der Internet-Telefonie

Die von vielen Marktforschern prognostizierten Umsatzverschiebungen vom konventionellen Telefonnetz zum Internet werden so sicherlich nicht stattfinden. Auch der Markt der konventionellen Telefonverbindungen ist in Bewegung geraten, wie wir seit dem 1.1.1998 in Deutschland wissen.

Die übersteigerten Erwartungen an Internet-Telefonie sind politischer Natur, wer vermag sicher zu sagen, ob übermorgen Internet-Telefonie nicht reguliert oder der Betrieb von ISDN-basierten Kompressions-Gateways lizenzfrei wird. Sicher bleibt, daß Internet-Telefonie eine beachtliches Integrationspotential besitzt, was sich zumindest in firmeneigenen IP-basierten Sprach-/Datennetzen (Intranet) und im "Surf 'n Call" für den Privatnutzer vortrefflich ausspielen läßt.

Voice over IP: Potential – Status - Trends

Werner Remmele

Siemens AG

Introduction

PBX technology has dominated voice communication for a long time. With the emerging of alternative solutions for voice transportation, e.g. using packet / cell oriented data networks for voice and video transportation as well, there is a chance for real convergence of the two technologies:

✦ Use of one single infrastructure for all transport.

✦ Use of one single SW infrastructure for PBX functionality as well as features and applications.

Moreover, the customer benefit of the convergence will be addressed by true integration of real-time communication in all applications thus providing a new technological solution for business systems for real-time communication.

Voice over IP (VoIP) comprises products in this emerging market of voice and data convergence. This market is addressed with business systems for real-time communication, thus providing PBX functionality as well as enhanced applications.

Thus, VoIP is the technology that enables the development of real-time communication on LAN based systems, without a PBX, but including terminals of all kind.

Technology

A VoIP system is based on an IP network that is implemented on top of a LAN[1]. Using this technology, the benefits of seamless communication through Internet / Intranet / Extranet can be used.

The connection to the existing PSTN or ISDN environment is done through a gateway, a component that carries out the necessary conversions (protocols mainly H.323 to H.320[2] as well as to other devices such as POTs or ISDN phones). The admission is controlled by the gatekeeper, that may reside inside the gateway itself or be implemented

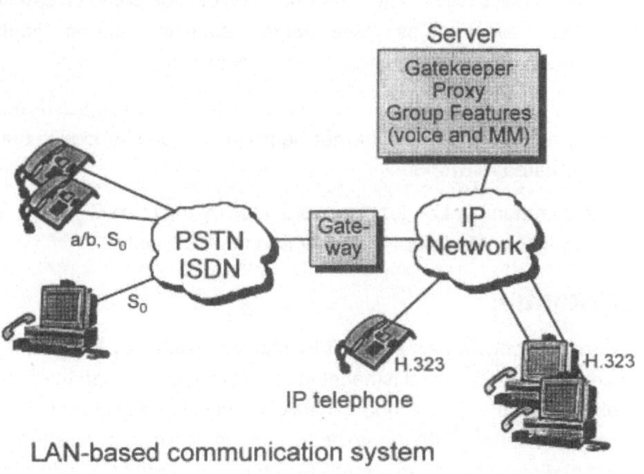

LAN-based communication system

---

[1]   Local Area Network. However, there will be implementations, where IP data will also be carried wireless.

[2]   H.323: Standard, that had originally been established for video-conferencing in LANs, but may also be used for voice only purposes. H.320 is the equivalent standard in the ISDN world.

on a server within the LAN.

Communication features will be provided on the feature server.

Calls can be set up from any station on the IP network (PC, NC or other terminals, that may be used) to any other station within the IP network or on the ISDN / PSTN, provided that the necessary functionality is available (e.g. video, data).

As packets on the IP network may carry data as well as voice or video data, no distinction between a pure voice call or a multi media call has to be made any more.

In order to provide real time communication over IP networks, the PBX functionality has to be provided with the alternative technology in a networked environment:

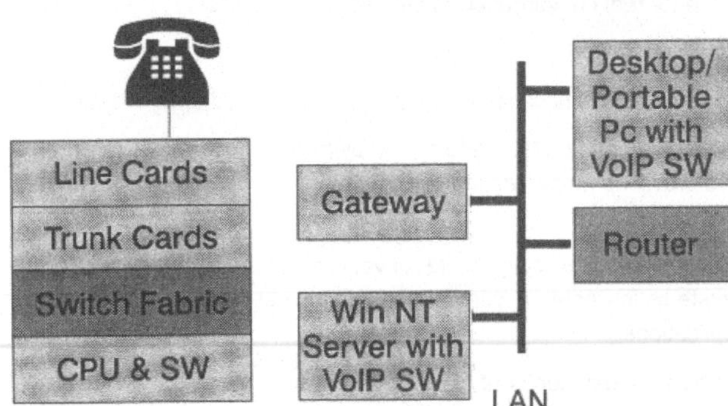

The function of the telephone is handled in the PC. In the case of digital phones on a PBX the Analog/Digital (A/D) Conversion happens in the telephone. In the PC the A/D conversion happens in the sound board or sound chip. The signaling functions of the line cards are carried out in the PC. The transmission function of the line cards happens in the LAN equipment.

The communication with the outside world which was handled by trunk cards in the PBX is handled by the H.323 Gateway for outside TDM circuits. The router handles outside communication with other native Internet Protocol clients. These could be offsite clients in the same Intranet, clients in a different site on the same Extranet, or clients anywhere in an IP network.

The Switch Fabric function of the PBX is performed by the IP network. This can happen in the LAN equipment for a single Ethernet Segment or it can happen in the router (or Ethernet Switches) in more complicated networks.

The common control functions provided by the main CPU and SW in a PBX will be distributed between the Windows NT Server and the PC clients.

Potential

Current technologies, e.g. CTI (computer / telephony integration) already offer functionality of similar kind. Here, the control is handled in a different way than the media stream. PBX technology still is needed for the media, and a intelligent functions are provided by computer that controls the PBX for some functions. With the unification of media and control streams on one single transport, the integration becomes much easier. Furtheron it offer the following advantages:

- Reduced cost for transportation (one infrastructure) as well as operational cost (e.g. one network management).

- Reduced cost for network management, as only one network for data and voice has to be managed.

- Lower costs of end devices due to 'data' oriented pricing.

- Increased efficiency in business processes due to real-time communication enabled data applications as well as easily adaptable feature sets.

✦ Re-engineering of business processes based on higher productivity.

✦ Faster and thus more efficient use of devices based on user friendly handling (point and click oriented user interface).

In future it will not be possible any more to provide only one of the core technologies within this field. Solution providers will have to be able to provide PBX like communication functionality as well as intelligent applications (horizontal and vertical). Therefore we shall provide a basic communication system, that uses in-house LAN technology as well as technology of our main partner 3Com for transport on the physical level. On top of this basic system there will be numerous applications, known from the PBX world as well as from the IT world. In order to be able to provide a complete product spectrum, the integration of VoIP functionality in any application running on standard based devices (PC, NC, or any Windows^{tm} based system) will be provided.

A complete product offering is shown here:

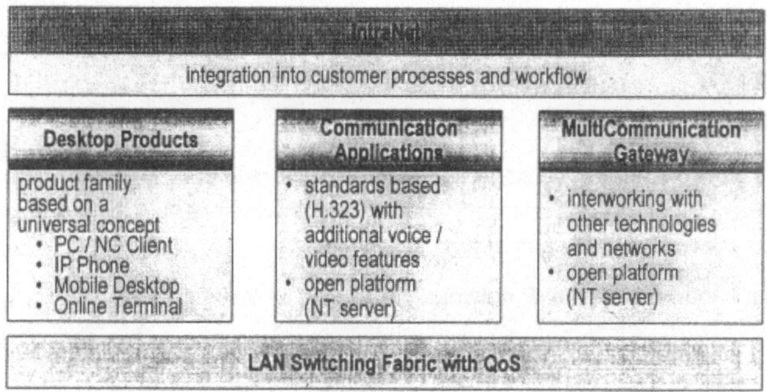

Benefits

The following chapter gives an overview on customer benefits, by using VoIP based solutions instead of others. We focus on answers to the following questions:

✦ Why change anything?

✦ Why choose MM over IP?

✦ Convergence of Voice and Data

✦ Multi-media communications

Why change anything?

There are many reasons for changing from the current existing PBX communication environment to or just adding a VoIP system. VoIP offers a variety of reasons for that change, as the following section shows:

One advantage is the reduction of life cycle and operational cost (cost of ownership). The change can be made because of any of these reasons:

✦ The company's intra-company telephone bill is too high

As soon as communication with other companies seems to become too expensive, the inherent possibility of using IP networks should be examined as an alternative.

✦ The cost of administering voice and data networks is too large

Administration of two networks is expensive as well as cumbersome. VoIP may provide unique management for example for moves / adds / changes (MAC's), independently of the terminal to be administered (data or real time communication). There is also only one single directory to be maintained.

✖ The employees do not communicate with each other as well as possible

Effective and efficient communication will be provided by the integration of real time voice, video and data into all applications, that are currently used at the site. Using the full potential of a converged technology the following possibilities will be available: Communication with the author or sender of a document is done by the click of the mouse, a discussion with the creator of notes in a document is started in the same way, as well as adding experts to an ongoing discussion. Even personalized secretarial functions, such as intelligent call forwarding, negotiated call back, call routing, etc. can be provided.

Why choose MM over IP?

VoIP solutions fit into any existing environment, so that there is actually no reason not to use VoIP's multi media functionality:

✖ If there is an IP network already:

If there is an IP network already - just use it. If the IP network is still the 'yellow cable', it will be upgraded anyway. Otherwise, as soon as 10 Mbs^{-1} are provided to the desktop, the basic requirement for adequate quality has already been achieved.

✖ If there is spare bandwidth on the IP network that should be leveraged

In many cases, the existing IP network offers spare bandwidth. Therefore using VoIP for real time communication will not overload the network.

✖ If there are bottlenecks in the IP network and an upgrade is necessary

In case of bottlenecks in the IP network, the upgrade will take place, anyway, independent of the use of VoIP.

✖ If a close interaction with customers and vendors via the Internet is necessary

Communication with the customers and vendors becomes an increasing requirement for efficient business. With its seamless integration into the Internet, VoIP provides all necessary interaction capabilities: voice, data, video.

Convergence of Voice and Data

The convergence of voice and data is provided on all levels: on the network as well as on applications and thus does not stop at the one outlet of a combined voice and data network. VoIP offers:

✖ Reduced Administration Labor Costs because of one network

  ✖ Only one set of user profiles to maintain

  There is only one directory to be maintained. This directory includes all relevant data for all sorts of communication.

  ✖ Only one administration system to learn and use

  There is only one administration system to be used. Therefore training costs as well as the cost of ownership is reduced.

✖ Wide area network cost reductions

  ✖ Combining voice and data traffic on a single link

There is only one single link to the outside world (in many cases, of course, there will both links remain for some time).

⬊ Increased voice compression

With the higher compression capabilities on data lines (that are currently not available on PBX's) communication cost is reduced.

⬊ Dynamic load balancing between voice and data

Wherever possible, a load balancing between voice and data can take place. Real time data (carrying voice, data or video) will be transported with higher priority.

The homogeneous technology allows to incorporate functionality much easier than in heterogeneous solutions. Such, novel support functions become possible, like communication aids (‚personal secretary') using intelligent agent technology to aid the communication

⬊ Reduce long distance communication costs

VoIP will provide means for reductions of long distance calls even over the existing 'least cost routing' techniques. With these techniques, all possible carriers are taken into account, including inter/intra/extranet, at a requested or calculated quality.

⬊ Increased effectiveness and efficiency of communication (e.g., find the right person the first time)

This function may be designed to create context sensitive help: for example, finding the right person to be connected to is a function of the system. The innovative system retrieves the necessary information out of different sources, e.g. the file currently being worked on, an e-mail or communication log, or even of the typical behavior of the user.

Multi-media communications

Many systems require dramatic changes to the business process involved. VoIP-based solutions will not force the user to change any program in use in order to make use of it ('you can use whatever you used to use....'). Moreover it becomes possible, to integrate the communication functionality into every application. The list gives some examples using real scenarios.

⬊ Call back to sender

⬊ Call person that made comment to MS Word Doc

⬊ Get advice from specialist based on application that is currently in use

As businesses rely on the productivity of their knowledge workers, it is essential for the success to be able to increase their productivity. VoIP based systems may support this goal by offering:

⬊ Clearer communication with multi-media

There is only one system for all means of communication. Video (still video or real time video) is just a natural extension of the system.

⬊ Get the right info everywhere, not to have to wait for input, such as faxes, e-mails, etc. when away from the desk

Whenever VoIP is used on portable devices (e.g. laptop computers), there is no need to use any other communication mechanism when the user is not at his / her desk. The request for a fax can be sent directly out of a meeting, and the fax will subsequently show up at the laptop computer in that meeting.

⬊ Change documents at any location, even during meetings, in order to increase the productivity of teams

Even existing documents can be directly changed during a meeting. This is a very powerful communication tool, as all participants of the meeting can be integrated into the communication and therefore directly accept changes, receive meeting minutes directly at the meeting's end, etc.

- All meeting participants see and agree to the result

- Less work after meeting

- Less chance for misunderstandings

- No uncertainty while waiting for meeting minutes

Status and Trends

VoIP systems have started penetrating the market about two years ago. Most of the solutions are still based on the telephony paradigm, thus providing telephone sets as the major communication device. Therefore these systems do not make use of the full potential of the technology.

All major PBX vendors are currently working on VoIP solutions, as well as many SW companies. This trend shows the dynamics of today's communication market: The traditional suppliers are threatened also by innovative SW companies – thus reflecting the true convergence not only on a technical level, but also on the product side.

Besides that, some problems have still to be resolved:

- The quality of VoIP solutions depends heavily on the underlying network. Before completely switching to such solutions, an analysis of the network has to take place. This analysis must cover the network components, the topology and the dynamic behavior.

- Solutions in VoIP are currently priced higher than pure voice solutions that are based on PBX's, due to the high cost of the gateway. This is more than compensated by the benefits through integration in the customer's workflow, but needs a careful analysis of the total life cycle cost.

- The reliability of PC's as end devices is not as high as for telephones. Therefore the system has to be based on reliable server platforms for mission critical installations.

Conclusion

The market for VoIP solutions is currently still in its infancy state, despite their undisputed advantages above traditional heterogeneous solutions. They still have to prove these advantages in day to day business environments.

With the rapid growth of products we believe, that the 'entry barrier' into this technology will be lowered so much, that within the next years they become the real alternative to traditional product offerings. Potential problems of interworking between different solutions have been covered by the rising and fully accepted standards. Other technology inherent problems, such as delay and QoS have been attacked and standard solutions will also be provided.

We are convinced, that the big advantages - like tailored solutions for every customer – are so convincing, that this technology will be widely accepted and used within the next years.

Werner Remmele
Siemens AG, PN VS LP 3
Hofmannstr. 51
D-81359 Munich
werner.remmele@pn.siemens.de

IP goes MultiMedia – mediaWays, a global IP-Carrier

Dieter Schinagel

Siemens AG

Eine schriftliche Zusammenfassung des Beitrages konnte nicht rechtzeitig fertiggestellt werden.

Oberweis
Modellierung und Ausführung von Workflows mit Petri-Netzen

Dieses Buch beschreibt einen auf Petri-Netzen basierenden integrierten Ansatz zur Modellierung von betrieblichen Abläufen und Objekten. Es wird eine evolutionäre Vorgehensweise zur Ablaufbeschreibung vorgestellt, welche von einer anwendungsnahen Notation zu einer präzisen und für die Ausführung mit Workflow-Managementsystemen geeigneten Notation führt.

Basierend auf den Sprachkonzepten für die Ablaufbeschreibung wird die Architektur eines Workflow-Managementsystems konzipiert.

Die Besonderheit dieses Systems besteht darin, daß Petri-Netze unmittelbar als Grundlage für die Ablaufkontrolle und -steuerung eingesetzt werden. Die beschriebenen Workflow-Engine kann auch als frei konfigurierbare Ablaufsteuerung von Standard-Softwaresystemen eingesetzt werden. Die Abläufe sind hier nicht »fest verdrahtet« in der Software enthalten, sondern als flexible Beschreibungen in Form eines Ablaufschemas gegeben, das im Rahmen der Ablaufmodellierung erstellt und optimiert worden ist.

Von Prof. Dr.
Andreas Oberweis
Universität Frankfurt/Main

1996. 302 Seiten mit
105 Bildern.
16,2 x 22,9 cm.
Kart. DM 54,–
ÖS 394,– / SFr 49,–
ISBN 3-8154-2600-6

(Teubner-Reihe
Wirtschaftsinformatik)

Preisänderungen vorbehalten.

B.G.Teubner Stuttgart · Leipzig

Schienmann
Objektorientierter Fachentwurf

**Ein terminologiebasierter
Ansatz für die Konstruktion
von Anwendungssystemen**

Das Buch zeigt, wie durch die Rekonstruktion der Terminologie eines Anwendungsbereichs das objektorientierte Fachkonzept für eine geplante Anwendung systematisch entwickelt werden kann. Der fachliche Entwurf ist dazu zunächst begriffsorientiert, anschließend objektorientiert ausgerichtet, wobei die in einer Rekonstruktionsphase ermittelten Fachbegriffe in der folgenden objektorientierten Spezifikationsphase zu Objekttypen mit ihren jeweiligen Merkmalen führen. Anhand eines durchgängigen Beispiels wird veranschaulicht, wie – ausgehend von der Untersuchung des Gebrauchs von Fachbegriffen im Anwendungsbereich – die Struktur und das Verhalten der gewünschten Anwendung entsprechend einem objektorientierten Spezifikationsrahmen festgelegt werden können.

Von Dr.
Bruno Schienmann
Informatik-Zentrum der
Sparkassenorganisation
(SIZ) Bonn

1997. 334 Seiten
mit 59 Bildern.
16,2 x 23,5 cm.
Kart. DM 69,80
ÖS 510,– / SFr 63,–
ISBN 3-8154-2305-8

(TEUBNER-TEXTE
zur Informatik, Bd. 20)

Preisänderungen vorbehalten.

B. G. Teubner Stuttgart · Leipzig

Berichte des German Chapter of the ACM

Band 33: **Ackermann/Ulich, Software-Ergonomie '91**
Fachtagung vom 18. bis 20. 3. 1991 in Zürich. 383 Seiten, DM 84,–/ÖS 613,–/SFr. 76,–

Band 34: **Friedrich/Rödiger, Computergestützte Gruppenarbeit (CSCW)**
Fachtagung vom 30. 9. bis 2. 10. 1991 in Bremen. 314 Seiten, DM 69,–/ÖS 504,–/SFr. 62,–

Band 35: **Hoffmann, Eiffel**
Fachtagung am 25./26. 5. 1992 in Darmstadt. 112 Seiten, DM 42,–/ÖS 307,–/SFr. 38,–

Band 36: **Schweiggert, Wirtschaftlichkeit von Software-Entwicklung und -Einsatz**
Fachtagung am 21./22. 9. 1992 in Ulm. 272 Seiten, DM 62,–/ÖS 453,–/SFr. 56,–

Band 37: **Ludewig/Schneider, Software Engineering im Unterricht der Hochschulen SEUH '92**
Workshop am 27./28. 2. 1992 in Stuttgart. 132 Seiten, DM 46,–/ÖS 336,–/SFr. 41,–

Band 38: **Raasch/Bassler, Software Engineering im Unterricht der Hochschulen SEUH '93**
Workshop am 25./26. 2. 1993 in Hamburg. 190 Seiten, DM 49,–/ÖS 358,–/SFr. 44,–

Band 39: **Rödiger, Software-Ergonomie '93**
Fachtagung vom 15. bis 17. 3. 1993 in Bremen. 330 Seiten, DM 78,–/ÖS 569,–/SFr. 70,–

Band 40: **Coy/Gorny/Kopp/Skarpelis, Menschengerechte Software als Wettbewerbsfaktor**
Arbeitstagung am 27./28. 1. 1993 in Bonn. 647 Seiten, DM 138,–/ÖS 1007,–/SFr. 124,–

Band 41: **Züllighoven/Altmann/Doberkat, Requirements Engineering '93: Prototyping**
Fachtagung vom 25. bis 27. 4. 1993 in Bonn. 383 Seiten, DM 88,–/ÖS 642,–/SFr. 79,–

Band 43: **Hußmann/Paech, Software Engineering im Unterricht der Hochschulen SEUH '94**
Workshop am 24./25. 2. 1994 in München. 178 Seiten, DM 54,–/ÖS 394,–/SFr. 49,–

Band 44: **Spillner/Breymann, Software Engineering im Unterricht der Hochschulen SEUH '95**
Workshop am 23./24. 2. 1995 in Bremen. 141 Seiten, DM 48,–/ÖS 350,–/SFr. 43,–

Band 45: **Böcker, Software-Ergonomie '95**
Fachtagung vom 20. bis 23. 2. 1995 in Darmstadt. 424 Seiten, DM 98,–/ÖS 715,–/SFr. 88,–

Band 46: **Daldrup, Menschengerechte Softwaregestaltung**
252 Seiten, DM 54,–/ÖS 394,–/SFr.49,–

Band 47: **Schweiggert/Stickel, Informationstechnik und Organisation**
Fachtagung am 28./29. 9.1995 in Ulm. 276 Seiten, DM 64,–/ÖS 467,–/SFr. 58,–

Band 48: **Forbrig/Riedewald, Software Engineering im Unterricht der Hochschulen SEUH '97**
Workshop am 27./28. 2.1997 in Rostock. 124 Seiten, DM 54,–/ÖS 394,–/SFr. 49,–

Band 49: **Liskowsky / Velichkovsky / Wünschmann, Software Ergonomie '97**
Fachtagung vom 3. bis 6. 3. 1997 in Dresden. 370 Seiten, DM 108,–/ÖS 788,–/SFr. 97,–

Band 50: **Sommer / Remmele / Klöckner, Interaktion im Web –
Innovative Kommunikationsformen**
Fachtagung am 12./13. Mai 1998 in Marburg. 221 Seiten. DM 68,–/ÖS 496,–/SFr. 61,–

B. G. Teubner Stuttgart